Conceptions of Space

Conceptions of Space

Beginning Geometries for College

William Hemmer

Grossmont College

Canfield Press  San Francisco

A Department of Harper & Row, Publishers, Inc.

New York   Evanston   London

CONCEPTIONS OF SPACE: Beginning Geometries for College

Copyright © 1973 by William Hemmer

International Standard Book Number: 0-06-383627-0

Library of Congress Catalog Card Number: 73-10192

   74 75 10 9 8 7 6 5 4 3 2

Library of Congress Cataloging in Publication Data

Hemmer, William, 1936-
   Conceptions of space.

   1. Geometry. I. Title.
QA453.H48   516   73-10192
ISBN 0-06-383627-0

to my father
Walter J. Hemmer

Preface

This book is intended for college students who have never studied geometry. It can be used in beginning geometry classes or in liberal arts math courses. No knowledge of classical algebra is needed except for the material in the appendices.

The material is divided evenly between Euclidean, Lobachevskian (hyperbolic), and Riemannian (elliptic) geometries. Each geometry is carried to the point where the sum of the angles of a triangle is investigated.

My purpose in writing this book was threefold. First, to teach geometry, a subject which has been part of the heritage of Western culture since its beginnings in pre-Christian Greece. Geometry is timeless and beautiful but this fact can no more be explained than can the beauty of a Mozart quartet. A teacher can help his students investigate, analyze, and probe, but in the end he cannot explain.

My second purpose was to exhibit the structure of mathematical systems by alternating axiom sets and investigating the outcomes. Using the three geometries, students with little mathematical training may be able to understand (and perhaps even to demonstrate) mathematical creativity.

My third purpose was to show that mathematics is the product of the human mind and as such is fundamentally an esthetic activity.

This book has several aspects which make it suitable for college students. The general tone is more philosophical than most geometry books intended for high school students. Proofs are treated as explanations in a more informal manner than the more common "Statement-Reason" boxes. Some research is required into the history of mathematics, a practice more suited to a college population.

Geometry as presented here is largely "mathematics for its own sake." For those who desire that their learning be useful and practical, some topics are included in the appendices which have applications outside the field.

Thanks to Lowell Cannon, Cuyahoga Community College, Richard Plagge, Highline Community College, Frank Denney, Chabot College, and Michael Mallen, Santa Barbara City College.

Special thanks to my colleague Larry Langley for helpful suggestions, and to Mrs. Ann Oliver for typing several versions including this one. And last, thanks to all the students who guided me patiently through it all.

William Hemmer
El Cajon, California
June, 1973

To the student

On rereading this book I find a great disparity between the richness geometry has in my mind and the sparseness of what I have actually written. This leads me to conclude that all I or anyone can give you are the bones. The flesh you must fill in for yourself.

This "fleshing in" will come with discussion and contemplation if it will come at all. Here are some suggestions which may breathe some life into the subject for you.

(a) Read the book carefully. You will quickly learn that you can't fruitfully skim-read a math book the way you might a history text. Almost every sentence counts. Omit a paragraph and something important is lost. Then remember what you have read.

(b) Ask yourself questions as you read. Some passages (especially the proofs of theorems) have to be read several times. Make notes in the margins. Be willing to turn back to something you have already read when advised to do so.

(c) Make lots and lots of freehand sketches. Make paper cutouts if necessary and if possible.

(d) Ask classmates and teacher lots of questions. Challenge anything and everything you don't understand (but try not to involve yourself in endless argumentation over unimportant minute details).

(e) Do every exercise, whether assigned or not.

Don't be afraid to be perplexed. While you are learning to make logical arguments, you will be wrong much of the time. This is normal. If you have more questions at the end of the book than you had at the beginning, the course will be successful.

The most important thing is to enjoy!

Contents

System 1   1

System 2   4

System 3   10
          Historical Background   14

System 4   EUCLIDEAN GEOMETRY   16
          On Changing Axiom Sets   47
          Transition   47

System 5   LOBACHEVSKIAN GEOMETRY   49
          Historical Background   69
          Transition   70

System 6   RIEMANNIAN GEOMETRY   71
          Conclusion   84

Appendices   86
          Appendix A   Proportions and Similar Triangles   86
          Appendix B   Square Roots and the Pythagorean Theorem   91
          Appendix C   Formulas   96
          Solutions to Exercises   97
          Notes on Uncompleted Proofs   108

Index   111

Conceptions of Space

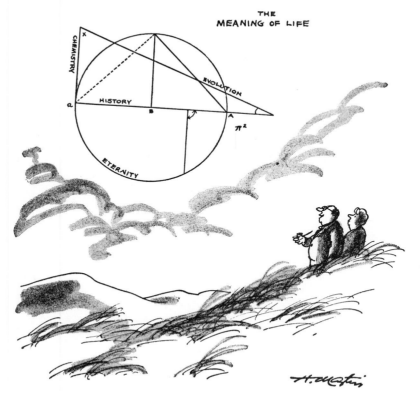

"I just wish I could see the faces of all my old students who kept asking me what good would their geometry be."

Cartoon by Henry R. Martin
Saturday Review, January 16, 1971

"We already know ... that the behaviour of measuring-rods and clocks is influenced by ... the distribution of matter. This in itself is sufficient to exclude the possibility of the exact validity of Euclidean geometry in our universe."

Albert Einstein

System 1

It is impossible to discuss the nature of logical arguments until we have made some. Thus, without further ado, think of a club divided into committees according to these rules:

Rule 1   Each <u>pair</u> of committees has exactly one (that is, at least one but no more than one) member in common.

Rule 2   Each <u>member</u> is on exactly two (that is, at least two but no more than two) committees.

Rule 3   There are exactly four committees.

Using only these three rules, answer the following questions:

Exercise 1-1   How many club members are there?

Exercise 1-2   How many members are on each committee?

To help you answer these questions, draw sketches, using rectangles to represent committees, and letters or numbers to represent members.

Do not continue to the next page until you have tried as best you can to answer these questions.

A picture drawn according to the rules of a system is called a
model of the system. System 1 has at least two models:

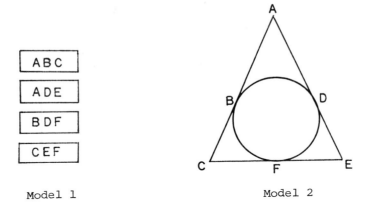

Model 1   Model 2

Model 1 uses rectangles for committees and letters for members. Model
2 suggests the possibility of using "points" and "lines" instead of
"members" and "committees."

Exercise 1-3   In the space below, rewrite the three rules of System 1,
               replacing the word "member" with the word "point" and the
               word "committee" with the words "straight line."

       An important question now arises. Is the line in model 2 which
contains the points B, D, and F a "straight" line? It may seem strange,
but the answer is yes! A model doesn't have to agree with our preconceived notions about straightness or in general with our past experience. A model of a system only has to agree with the rules of that
system! We will return to this idea later. (Models 1 and 2 above are
not the only possible models of this system.)

       An existence rule for a system is a rule that states uncondition-
ally that an object exists.

Exercise 1-4   Only one of the rules of System 1 is an existence rule.
               Which one?

When using the rules, the best sequence is to find the existence rule(s), then look for those rules whose subject (in the grammatical sense) is the same as the "things" that the existence rule says exist. For example in this system, start with rule 3, then use rule 1. (I take the subject of a sentence in the broad sense, that is, the subject is everything that comes before the verb and the predicate everything that comes after. This is true often enough to be convenient. Grammarians, forgive!)

We have two models of System 1 (or perhaps two versions of the same model) with a total of six members--that is, three members on each committee. How can we be sure that there isn't another model with more than six members which also fits the rules of System 1? The following exercises attempt to deal with this question.

Exercise 1-5   Explain why there cannot be just five members. (Hint: Eliminate someone from model 1, then see what happens.)

Exercise 1-6   Explain why there cannot be seven members. (Hint: Add someone to model 1, then see what happens.)

You have just been studying a <u>deductive system</u>. A deductive system is a set of basic rules (called axioms) and the collection of all statements which can be deduced from the basic rules.

Exercise 1-7   List some statements which are true in System 1 but which are not axioms.

Let's look now at the basic rules (axioms) of another system. When you start with a new set of rules, you ignore the old set (System 1) and start off fresh.

AXIOMS OR RULES

1. Take <u>any</u> pair of members. Then there is exactly one (that is, at least one but no more than one) committee containing <u>both</u> of them

2. Any two committees have at least one member in common.

3. There is at least one committee.

4. Every committee contains exactly three members.

5. Not all members serve on the same committee.

Do not try to grasp the entire system or to form a mental picture of its model before continuing. The meaning of the axioms and a model of the system will emerge gradually as you work with the system. Just read the axioms several times, then begin the exercises. The exercises are designed to help you understand the axioms.

Exercise 2-1   Which axiom or axioms above is an existence axiom?

Exercise 2-2   Explain why there must be at least three members. (Use a sketch.)

Exercise 2-3   Explain why there must be at least four members and at least two committees. (Start with the fact that there are at least three.)

Exercise 2-4   Explain why there must be at least five members.   (Start
               from exercise 2-3, that is, that there are at least four
               members.)

Exercise 2-5   Redraw the model for System 1 below and explain why it
               does not work for System 2.

Exercise 2-6   Using circles or rectangles for committees and letters or
               numbers for members, build a model for System 2.   (This
               is a tough one!   Begin with axiom 3, then use 4, 5, and 1
               in that order.)

Exercise 2-7   Does axiom 1 make axiom 3 unnecessary; that is, if axiom
               1 is true, doesn't axiom 3 have to be true?

Exercise 2-8   Does axiom 1 make axiom 2 unnecessary?

Exercise 2-9   In the space below, recopy the axioms of System 2 replacing the word "member" with the word "point" and the word "committee" with the word "line."

According to the "new" axiom 3, there is at least one line, so we can make the following sketch:

The "new" axiom 4 says that the line contains three points, so we can add three points to the sketch, like this:

Exercise 2-10 Continue adding to the sketch above using the "new" axioms 5 and 1, and explain why there must be at least four lines and at least seven points.

Exercise 2-11 Construct a "point-line" model for System 2.

ON BUILDING A SYSTEM

An axiom (or rule) is a statement which is accepted as true just for the sake of progress. One never attempts to prove an axiom. Instead, its truth is assumed for the sake of the argument. Such statements are also called "postulates" or "assumptions."

An axiom set is a collection of two (2) or more axioms all of which are assumed to be true at the same time.

A statement which is true in a particular system but which is not an axiom is called a theorem. The word "true" in this context means that it has been "worked out" (explained, defended, proved) using the axioms as a foundation or starting point. A theorem then, is a statement the truth of which must be explained or upheld using axioms, definitions, or other theorems which have already been proved. (A definition is a special kind of axiom. More on this later.)

A proof of a theorem is an explanation or argument showing how the statement follows logically from axioms or from earlier statements which have already been proved. Complete exercises 2-12 through 2-15.

Exercise 2-12 According to the above, a proof is a kind of explanation. Contrast a proof with the kind of explanation necessary to explain why there is smog in the Los Angeles Basin.

Exercise 2-13 State at least two of the theorems of System 1.

Exercise 2-14 State at least three theorems of System 2.

Exercise 2-15 Prove that in System 2 there are at least four members and at least two committees. (Hint: Look at exercise 2-3, page 4)

MORE ON BUILDING A SYSTEM

Recall that a deductive system is the entire collection of axioms (the axiom set), all theorems deduced from the axioms, and all versions of the model.

Exercise 2-16 How many deductive systems have you studied so far in this course?

The objects of a system are the "things" that the system deals with. In geometry, the objects of the systems are usually points, lines, and planes.

The relations of a system are descriptions of how objects are located with respect to each other. For instance, three points can be related to each other by saying that one lies "between" the other two. A point can be related to a line by saying that the point "lies on" the line. The same situation can be described by saying that the line "goes through" the point or that the line "contains" the point.

A meta-axiom is a statement which is assumed to be true about all axiom sets or about all deductive systems and their models.

Meta-Axiom 1   (Axiom of Innocence)   In each deductive system, the axio[m]
and the theorems deducible from the axioms comprise the
total of your knowledge about the objects of the system
about the relations between or among the objects. (In
short, you must pretend that all you know about the syst[em]
is what the axioms say and what can be deduced from them[.]

For centuries, oriental philosophers have begun the instruction o[f]
their pupils by urging them to rid their minds of whatever they though[t]
they knew in order to approach the world innocent of illusion. "To th[e]
man who doesn't pretend that he knows anything, the world will be a c[on]
stant source of surprise and delight." This idea of clearing your min[d]
of preconceived notions before you begin to learn is embodied in meta-
axiom 1.

In this book the words "argument," "proof," "explanation," "demo[n]
stration," and "deduction" all mean the same thing. So do the words
"all," "each," "every," and "any." And the phrase "exactly x" always
means "at least x but no more than x ."

Exercise 2-17   Is the statement immediately above another meta-axiom?

Exercise 2-18   There are some earlier statements which are really met[a]
axioms but haven't been called such. What are they?

Exercise 2-19   Does meta-axiom 1 say that you cannot use any outside
knowledge when you're working in a system?

Exercise 2-20   Could there be such a thing as a meta-meta-axiom?

Exercise 2-21   If an axiom says something about "exactly three" point[s]
might it possibly mean two points? Might it possibly
mean four?

A word of advice: Many students wait till a test is announced before they decide to assimilate an idea or argument, to "really" learn it. If you are somewhat mystified by geometry at this point, ask yourself if you are doing this. If so, begin to learn the ideas as though you were going to be tested on them every day. This takes more effort, but it will raise you to a higher plane as a learner. (No pun intended.)

We proceed now to a third deductive system.

AXIOMS

1. If there are two points on a line (call them A and B), then there is no other line through <u>both</u> A and B.

2. If L represents any line and if there is some point (call it P) which is not on L, then there is exactly one other line through P which has no points in common with L.

3. There is at least one line.

4. Every line contains exactly three points.

5. Not all points are on the same line.

6. If L is any line and P is any point not on L, then there is exactly one point (call it Q) on L which is not joined to P with a line; that is, there is exactly one point on L which is not on a line with P.

Make sketches when possible to help you do the following exercises

Exercise 3-1 Which axiom or axioms are existence axioms?

Exercise 3-2 Must every <u>pair</u> of points have a line on them both? (Must some line be through every pair of points?)

Exercise 3-3 Why does(n't) the model for System 2 also serve as a model for System 3?

Exercise 3-4 Prove the theorem: There are at least four lines.

Every simple statement has a denial. For example, the statement "It is raining" is denied by saying "It is not raining." The statement "Two line segments are not equal" is denied by saying "Two line segments are equal."*

Exercise 3-5 Give the denial of each of the following statements. Use the answers to help you by checking each answer as you go.
(a) Angle A is acute.
(b) Line L passes through point P.
(c) A line does not exist
(d) Line L_1 is not perpendicular to line L_2.

Meta-Axiom 2 For every simple statement, either the statement is true or its denial is true. There are no other possibilities. If a statement is true, its denial is false.

Exercise 3-6 Would the meaning or content of meta-axiom 2 be changed if either of the following things were done?
(a) The last sentence were dropped.
(b) The last sentence were changed to: If a statement is true, its denial is false, and if the denial is true, the statement is false.

INDIRECT PROOF

An indirect proof begins with the denial of what you are trying to conclude; that is, by supposing that the thing you want to conclude is not so and then trying to find a contradiction of something you already know is so, like an axiom or another theorem. Once you find a contradiction, the thing you "supposed" (the denial that led to the contradiction) must be false, and so the conclusion must be true.

For example, a theorem of System 1 could be: "There cannot be more than six members." You have already proved that this statement is true. The method you used was to say "Suppose there were more than six members,..." and then to explain why a seventh member leads to a contradiction of axiom 1.

Exercise 3-7 In which exercises up to now have you used an indirect proof? (Take a few minutes to review the earlier exercises.)

This cursory discussion of denials barely scratches the surface of a complex subject, though it is sufficient for the purposes of the work in this text. For a more thorough discussion of the analysis of propositions, see An Introduction to Logic by Morris R. Cohen and Ernest Nagel (New York: Harcourt, Brace & World, Inc., Harbinger Edition, 1962).

Here is an example of an indirect proof in System 3:

Theorem 3A   Two lines cannot meet in more than one point.

Proof: Step 1:   Deny what you're trying to prove, that is, suppose t[
what you want to prove is false.
        In this example, suppose that two lines can meet in more than on[
point.   (This statement is called "the supposition" for obvious reaso[

Step 2:   Draw a picture of what you are "supposing."
          For this example:

Step 3:   Search the axioms and previously proved theorems for an expl[
nation of why the picture (and thus the "supposition") is impossible.
        In this example, axiom 1 says that exactly one line can pass
through both A and B though in the picture there are two lines throug[
both.   Therefore the picture must be impossible and the supposition
which led to it must be false.   Then if the "supposition" is false, t[
theorem must be true!

        Do not continue until you have read this proof several times to
the steps in your mind.
        Now use the three steps just given to prove the following theore[
(The proof is given with the solutions so you can check yourself, but
don't look till you've given it a good try.)

Theorem 3B   There is no point that all lines of the system lie on.
            (That is, there is no point that all lines pass through.)

12

A LITTLE LOGIC

(1)   If it rains, then we get wet.

(2)   If x is a man, then x is mortal.

(3)   If two parallel lines are both intersected by a third line,
      then the alternate interior angles are equal.

The three statements above are examples of a type of statement
called an <u>implication</u>. We could rewrite them this way:

(1)   It rains implies we get wet.

(2)   x is a man implies x is mortal.

(3)   Two parallel lines being cut by a third line implies that the
      alternate interior angles are equal.

Implications always have the form "If..., then..." The part which
follows the word "If" is called the <u>hypothesis</u>, the part which follows
the word "then" is called the <u>conclusion</u>. (Fix these things firmly in
your mind because they will be used repeatedly throughout the book.)
Many <u>theorems</u> will be implications or will be capable of being re-
written as implications.

Exercise 3-8   Mentally determine the hypothesis and conclusion of each
               of the three statements above.

Exercise 3-9   Is it possible for the statement: "If dogs are cats,
               then horses are pigs" to be true? Try to answer this
               question before continuing.

To "prove" an implication does <u>not</u> mean to prove the truth of
either the hypothesis or the conclusion separately. Instead, it means
to assume the truth of the hypothesis and then use the axioms, the
previously proved theorems, and the definitions to explain why the
conclusion has to be so. The implication in exercise 3-9 could be considered true.* "Truth" in this case does not depend on either the hypothesis or the conclusion being true, but only on whether or not the
truth of the hypothesis forces the conclusion to be true. To say that
an <u>implication</u> is true is another way of saying that a "force" exists
between the hypothesis and the conclusion which is of such a nature
that the truth of the hypothesis sets "in motion" a mechanism which
causes the conclusion to be true. The description of this mechanism
is the proof.

_____

* It is an axiom of logic (and hence of mathematics) that if the
hypothesis of an implication is false, the entire implication is true,
regardless of the conclusion. This curious idea often leads to interesting games in logic, but has no application in this book.

The majority of students who study geometry do not fully understar
the axioms and theorems the first time they read them. If this is the
case with you, it should not cause you any undue consternation. Understanding usually comes with discussion, rereading, making sketches, and
contemplation. Very few people understand mathematical ideas instantly
Like all growth processes, it takes time.

HISTORICAL BACKGROUND

You are about to begin the study of Euclidean geometry. Before yo
do, it seems a good idea to do a little research in order to obtain sor
perspective on geometry, that is, to see geometry as the product of the
human mind, to see it in its human context.

It is in this spirit that the following assignment is made. Perh.
in a small way this assignment will help you get a whisper of the mist
beginnings of human thought.

Read <u>at least three</u> of the following works or similar works which
your teacher will assign. This assignment, properly done, will mean a
least two hours of reading.

The questions at the end are for you to use to help yourself stru
ture your reading. It is not intended that you answer the questions a
then quit. They are only a guide. Read the questions several times b
fore beginning to read the assigned works. Reading with questions in
mind is always more fruitful than random perusing. Hopefully other qu
tions will arise in your mind as you read.

Don't spend a lot of time trying to understand the <u>mathematical</u>
explanations you run across. Just skip over them and keep reading.

(1) <u>Science Awakening</u>, B.L. van der Waerden (Oxford: Oxford Universit
    Press, 1961). Pages 15, 30, 35, 37, 76, 82-95, 100, 125, 138, 14
    195, 201, 208, 227, 264, and plate 19

(2) <u>Mathematics In Western Culture</u>, Morris Kline (Oxford: Oxford
    University Press, 1953). Chapters IV, X; Plates IX, X, XIII, XIV
    XV, XVIII, XX-XXV

(3) <u>The Role of Mathematics in the Rise of Science</u>, Salomon Bochner
    (Princeton: Princeton University Press, 1966). Pages 20-28, 30-3
    48-58

(4) <u>Evolution of Mathematical Concepts</u>, Raymond L. Wilder (New York:
    John Wiley & Sons, 1968). Pages 79-89

(5) <u>An Introduction to the Foundations and Fundamental Concepts of
    Mathematics</u>, Howard Eves and Carroll Newsom (New York: Holt,
    Rinehart & Winston, 1965). Pages 1-14, 30-51

(6) <u>The Search For Truth</u>, Eric T. Bell (New York: Reynal & Hitchcock
    1934). Pages 43-60, 128-139

(7) World of Mathematics, vol. 1, James R. Newman (New York: Simon & Schuster, 1956). Pages 79-86, 100-103, 176-177, 179-185, 402-416

(8) Mathematics in the Making, Lancelot Hogben (Garden City: Doubleday & Co., 1960). Pages 50-57, 74-84, 98-103

(9) Makers of Mathematics, Alfred Hooper (New York: Random House, 1948). Chapter 2

(10) An Introduction to the History of Mathematics, Howard Eves (New York: Holt, Rinehart & Winston, 1960). Chapters 3, 4, 5

(11) A History of Mathematics, Carl B. Boyer (New York: John Wiley & Sons, 1968). Chapters 5, 6, 7

(12) Development of Mathematics, Eric T. Bell (New York: McGraw-Hill, 1945). Chapter 3

(13) Mathematics--A Cultural Approach, Morris Kline (Reading: Addison-Wesley, 1962). Chapters 2, 3

(14) Mathematical Thought From Ancient to Modern Times, Morris Kline (Oxford: Oxford University Press, 1972). Chapters 3, 4

(15) History of Mathematics, vol. 1, David E. Smith (New York: Dover, 1951). Chapters III, IV

Some questions to keep in mind as you read:

1. What event in nature is said to have given rise to the development of Egyptian geometry?

2. How did the geometry of the Egyptians differ from that of the Greeks and why was it different?

3. What is the source of most of our knowledge of Egyptian math?

4. What culture, in addition to Egyptian, greatly influenced the Greeks in their development of mathematics?

5. Is mathematics a discovery or an invention?

6. Four names that are "big" in the history of mathematics are Thales, Pythagoras, Euclid, and Archimedes. What did these people do that they should be remembered today?

7. How much truth is there to the statement: The ancient Greeks worked solely with geometry and never with algebra?

System 4

euclidean geometry

The fourth deductive system in this book is that compiled by the Greek mathematician Euclid (c. 300 B.C.). The first group of axioms for this system, modeled after the revisions of David Hilbert, begin below.

Since the words "point," "line," and "plane" used below have not been defined, you are free to use any mental image you please, as long as the axioms are not violated. A point may be thought of as a glimmer in space or, more materialistically, something like an atom but smaller. A line may be thought of as the path of a ray of light. A plane may be thought of as a flat surface, exceedingly thin. These are useful mental images to help in visualizing the geometry though they are not logically necessary.

Meta-Axiom 3   When several models (or several versions of a model) are correct for a deductive system, always use the model (or version) which agrees most with your past experience. If several agree with your past experience, use the one that is simplest.

AXIOMS OF CONNECTION

1. For any two distinct points, there is exactly one line through them both.

2. There are at least two points on every line.

3. There are at least three points on every plane, not all of which are on the same line.

4. For any (every) three distinct points that do not lie on the same line, there is exactly one plane on all three.

5. If two points of a line lie on a plane, then the entire line lies on that same plane.

6. There are at least four points, not all of which lie on the same plane.

7. If two planes have one point in common, then they have exactly
   one line in common.

   There are more axioms for Euclidean geometry. The Axioms of Connection are the first group. Read them several times to fix them in
your mind, then begin the exercises. The exercises are designed to
bring out the meaning of the axioms. Begin by proving the following
theorem. (Make sketches.)

<u>Theorem 4A</u> Two distinct lines cannot have more than one point in common.
             (That is, two lines cannot meet in more than one point.)

Exercise 4-1 Only one of the Axioms of Connection is an existence axiom,
              the kind of axiom you need in order to begin building a
              model. Which one is it?

Exercise 4-2 Explain why there is at least one line and at least one
              plane.

Exercise 4-3 Explain why there are at least 6 lines. (Hint: Use axioms
              6, 3, and 1.)

Exercise 4-4 Which of the three earlier systems has an axiom like
              Axiom 1 of Connection?

Exercise 4-5 On the basis of the Axioms of Connection alone, is it
              possible to prove that a line contains more than two
              points?

Exercise 4-6 True or false or neither: Lines are infinite in extent.

Exercise 4-7 Which axiom says, in effect, that a line cannot "jump off"
              a plane?

<u>Theorem 4B</u>   If two lines intersect, then there is exactly one plane
           which contains them both.   (Hint: See axioms 2, 4, and 5.)

<u>Proof</u>:   We begin with two intersecting lines,
Call them L_1 and L_2, and the point of inter-

section R, like this:

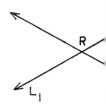

By axiom 2, both L_1 and L_2 have at least
one other point.   Label these other points
A and B, like this:

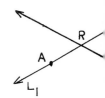

Points A, R, and B are not all on the
same line.   Now use axioms 4 and 5 to
finish the proof.

       Before continuing, review what is meant by the "objects" and
"relations" of a system, page 7.

<u>Meta-Axiom 4</u>   (Axiom of Completeness)   Given any statement about the
           relations between the objects of a system, it should be
           possible to tell on the basis of the axioms of the system
           whether the statement is true or false.

Exercise 4-8   Do the Axioms of Connection alone comprise a "complete"
           deductive system?   (Hint: Look at exercises 4-5 and 4-6

Exercise 4-9   There's nothing in the axioms that says that a line cann
           look either like this

           or like this

           You will most likely use the second version, however, whe
           making sketches.   Why?   (Hint: See Axiom 7 of Connectio

Unlike the three earlier systems, Euclidean geometry concerns itself with the relative sizes of things. As an aid to thinking, consider the word "congruent" to mean "same size." Two objects can be thought of as congruent if they can be made to fit precisely over each other in some way, with no overlap. This description is only to aid in visualizing the ideas described below. It is not a proper mathematical definition.

AXIOMS OF ORDER

1. For any two points of a line, there exists at least one third point of the line which lies between them.

2. For any three points on a line, exactly one lies between the other two.

Study the following definitions carefully. Make sketches.

Any two points (call them A and B) of a line together with all the points between them comprise a <u>segment</u>. The two points you started with are the <u>endpoints</u> of the segment. The segment is named by its endpoints, for instance, segment AB or segment BA. The <u>midpoint</u> of a segment (call it M) is a point between the endpoints of the segment that is located in such a way that AM is congruent to MB.

From any three points A, B, and C not all on the same line, we immediately get three segments: segment AB, segment AC, and **segment BC**. The three segments together comprise what is called a <u>triangle</u>. The segments are called the <u>sides</u> of the triangle. Each of the three points is called a <u>vertex</u>. Each vertex lies on exactly two sides. The side which does <u>not</u> contain a particular vertex is said to be <u>opposite</u> that vertex. The triangle is named by the letters at its vertices, for example, triangle ABC.

3. Every segment has exactly one midpoint.

4. If a line and a triangle are in the same plane and if the line inter-sects a side of the triangle at a point which is not a vertex, then the line also intersects one of the other sides of the triangle.

Make sketches to help you do the following exercises.

Exercise 4-10   Use Axiom 2 of Connection and Axiom 1 of Order to explain why there are an infinite number of points on every line.

19

Exercise 4-11   What would be the result if the phrase "at a point which
                is not a vertex" were dropped from Axiom 4 of Order?

Exercise 4-12   Does Axiom 4 of Order allow the possibility that a line
                might intersect all three sides of a triangle?

Exercise 4-13   Use Axiom 2 of Order to explain why a line cannot close

                on itself like this   ∞   or like this   ○

                (Hint: Use an indirect argument.)

Exercise 4-14   Draw a triangle, label its vertices A, B, and C, then tel
                which sides are opposite which vertices.

DEDEKIND'S AXIOM

    If all points on a line are separated into two nonempty sets L and
R in such a way that every point of L is to the left of every point in
R, then there is exactly one point T which is the boundary of this separation and T is either the rightmost point of L or the leftmost point
of R.

    To get a "feel" for Dedekind's Axiom, try imagining two points both
of which have the property described.

    Dedekind's Axiom fills the line with points. It is the explicit
admission of our inability to imagine a space or region without points,
a region where there is truly nothing.
    For an analogy, a point "does to" a line what each instant does to
time, namely, divides it into two parts which never overlap, past and
future. We think of time as the endless succession of such instants,
each following the next in a smooth flow without breaks or jumps. (Even
though we cannot remember each instant of the last hour, our imagination
or intuition assures us that that hour was completely filled with
instants, each of which divided time into past and future. Thus our
minds fill in instants whose existence our memory cannot verify.)

From Dedekind's Axiom, we have lines which are continuous, without gaps.*

AXIOM OF SEPARATION

Every line partitions every plane in which it lies into two half-planes called the left half-plane and the right half-plane (or the upper and lower half-planes) in such a way that:

(1) Exactly one of the following is true: every point of the plane is in the left half-plane or in the right half-plane or on the line.

(2) For any point in one half-plane and any point in the other, the line joining them intersects that line which formed the half-planes.

(3) If two points of a line are in the same half-plane, then every point of the line between them is in that half-plane.

Use the above axiom to do exercises 4-15 and 4-16.

Exercise 4-15 Explain why it is not possible for two lines (call them L_1 and L_2) to intersect in the manner illustrated, that is, to intersect without "crossing over." (Hint: See part (3) of the Axiom of Separation.)

Exercise 4-16 Explain why, if two lines are in the same plane (call them L_1 and L_2, it is not possible for L_1 to make a kind of "end run" around L_2 (as sketched on the right) without the lines intersecting. (Hint: A and B are points of L_1. How many lines can pass through both A and B? See part (2) of the Axiom of Separation.)

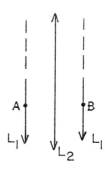

There is still no compulsion to picture a straight line as the familiar "straight" of our past experience. However, the Axiom of Separation makes it increasingly difficult to picture anything else.

---

* For further readings on this subject, see Chapter Nine of Number: The Language of Science by Tobias Dantzig (New York: The Macmillan Company, 1954)--a most marvelous book.

New objects or relations are introduced into a deductive system by means of statements called <u>definitions</u>. (There have already been some definitions on page 19.) Why introduce new objects and relations into a system? Why not just "make do" with those you have? The reason, like other fundamental reasons in math, is alogical. The interesting things that can be said about the old objects and relations may have been said or maybe one just feels an irresistible urge! The most fundamental desire of mathematicians is to continue the game, that is, to continue "mathematizing." If we don't introduce new objects and relations, the game may end!

You should know these three definitions.

(1) <u>Ray</u>: Any point on a line divides the line into two parts which have only point P in common and which together make up the whole line. Each part is called a <u>ray</u>. A ray is named by the letter at its endpoint and a letter at some other point on the ray. The two rays on the right would be named ray PA and ray PB.

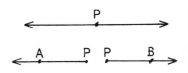

(2) <u>Angle</u>: If two rays are joined at their endpoints, the resulting figure is called an <u>angle</u>. The rays are called the <u>sides</u> of the angle. The common point is called the <u>vertex</u>. An angle is named using three letters; one letter at a point on one side, one letter at the vertex, and one letter at a point on the second side. This is illustrated by the three figures on the right. The letter at the vertex is always named in the middle.

However, when there is no chance of confusion, the letter at the vertex can be used alone to name the angle and sometimes a number written by the vertex is used. (Use the method that is most convenient.)

Notice in the third figure that the single letter O would not be used to name any of the angles because there are three angles with vertex at O.

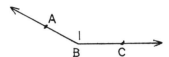

Angle ABC (or angle CBA) or simply angle B or angle 1.

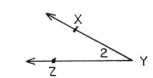

Angle XYZ (or angle ZYX or angle Y or angle 2)

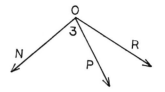

Angle NOP (or angle 3)
Angle NOR and angle POR

(3) <u>Interior and exterior of an angle</u>: Because of the Axiom of
Separation, an angle divides the plane in which it lies into two
regions in such a way that if a segment is drawn from any point in
one region to any point in the other, the segment will intersect
the angle. For example, in the figure below, points A and B are in
one region, points C, D, and E are in the other. (The two regions
are formed by angle XYZ.)

    Notice that if points A and B were the endpoints of a line seg-
ment, that line segment would not intersect the angle. The same is
true of points C and E. However, if points C and D were the endpoints of a segment, that segment <u>would</u> intersect the angle. Convince yourself that this is true before continuing.

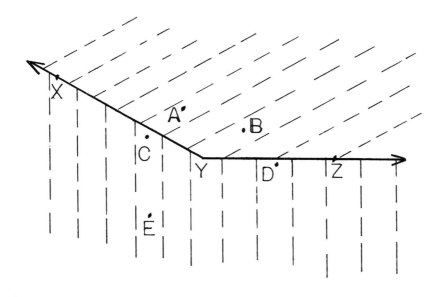

    One of the two regions is called the <u>interior</u> of the angle and
one is called the <u>exterior</u> of the angle. It is going to be important in later work to be able to tell the difference, so here is
how you tell! One of the two regions in the picture (the region of
A and B) has the property that if <u>any</u> two points in the region are
the endpoints of a line segment, that line segment will not intersect the angle. This region is called the <u>interior of the angle</u>.
The region which doesn't have this property is called the <u>exterior
of the angle</u>. The angle itself is in neither region.

Exercise 4-17   Name the points in the interior and the points in the
exterior of the following angles:

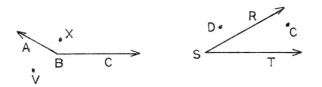

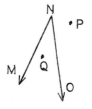

Exercise 4-18   Suppose L_1 and L_2 are two
lines which intersect at point
O, as shown, and that P and Q
are two other points on L_1 and
R and S two other points on L_2.
Now suppose L_3 is a third line
which lies in the interior of
angle POR and which passes
through O.   Use the Axiom of
Separation to explain why L_3
must also lie in the interior
of angle QOS.

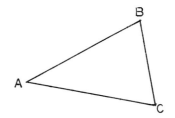

Now to change the subject a bit...

The triangle on the right has six
parts.   Try to name them before continuing.

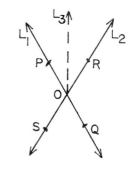

The six parts are the three sides: AB, BC, and CA, and the three
angles: angle A, angle B, and angle C.   Angle A is called the <u>included</u>
<u>angle</u> between sides AB and CA.   (Notice that the letter of the angle
(A) is the letter that the two sides have in common.)   Angle B is
included between AB and BC, and angle C is included between CA and BC.
Similarly, side AB is included between angle A and angle B, etc.

Exercise 4-19   True or false: In a triangle, every angle is included
between two sides and every side is included between
two angles.

Two <u>triangles</u> <u>are</u> <u>congruent</u> if all six parts of one are congruent respectively to the six parts of the other.   In a sketch congruent

segments are marked like this ⊥⊥⊥ or like this ⫽⫽

Congruent angles are marked like this

or like this

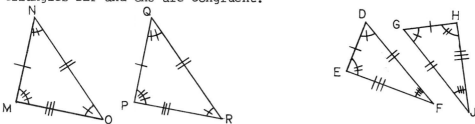

   In the figures below, triangles MNO and PQR are congruent, and triangles DEF and GHJ are congruent.

Exercise 4-20   Fill in the blanks, referring to the figure above.

Side DE is congruent to side _____,

side QR is congruent to side _____,

angle R is congruent to angle _____,

angle H is congruent to angle _____.

(The purpose of this exercise is to give you a feel for congruent triangles. If you still feel in doubt, try sketching several pairs of congruent triangles, marking the six corresponding congruent parts, then check them with a teacher.)
   As an aid to intuition, think of congruent figures as figures which fit precisely over each other, with no overlap.

AXIOMS OF CONGRUENCE

1.   If AB is a segment of a line L and C is a third point on the same or some other line, then there is on each side of C and on the same line exactly one fourth point D which will make AB congruent to CD.

2.   Every segment is congruent to itself. If any segment is congruent to a second segment, then the second segment is congruent to the first.

3.   If a segment (call it seg AB) is congruent to another segment (call it seg CD) and seg CD is congruent to a third segment (call it seg EF), then seg AB is congruent to seg EF. (This is a property of congruence called <u>transitivity</u> .)

<u>Definition</u>: If O is a point on a line L and
if OR is a ray or segment with its endpoint
at O, and if OR and L form two congruent
angles, then L and OR are said to be
<u>perpendicular</u> and the angles they form are
called <u>right angles</u>.

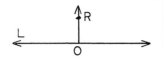

4. Axiom of Perpendicularity. If L is any line and P is any point,
    then there is exactly one line through P which is perpendicular to L

5. Suppose ABC is any angle. If O is a
    point on some line L, and M is a
    second point on L, then (on either
    side of L) there is exactly one ray
    with its endpoint at O which forms
    an angle with ray OM which is
    congruent to angle ABC.

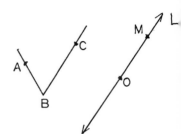

6. Every angle is congruent to itself. If any angle is congruent to a
    second angle, then the second angle is congruent to the first.

7. If an angle (call it ABC) is congruent to a second angle (call it
    POR) and angle POR is congruent to a third angle (call it XYZ),
    then angle ABC is congruent to angle XYZ. (Transitivity again, this
    time for angles.)

8. If there are two triangles (call them ABC and XYZ) and if AB is con-
    gruent to XY and angle ABC is congruent to angle XYZ and BC is con-
    gruent to YZ, then triangle ABC is congruent to triangle XYZ.

    (Axiom 8 can also be stated: If two sides and the included angle of
    one triangle are congruent respectively to two sides and the include
    angle of another, then the triangles are congruent.)

    Read these axioms over several times before continuing.

        The last axiom is very important. Even though a pair of congruent
triangles means that six parts of one are congruent to six parts of the
other, axiom 8 says that if you know something about only three of the
six parts (namely, two sides and the included angle of one congruent to
two sides and the included angle of the other) then the other two angles
and the remaining side of one triangle must also be congruent to the
corresponding angles and side of the other.

Exercise 4-21   Below are six statements which are paraphrases of the
                Axioms of Congruence. The paraphrases are in less pre-
                cise, more "everyday" language. For each statement, tell
                which axiom is being paraphrased.

                (a) If you have an angle and you want to draw another
                     angle somewhere which is congruent to it, you may
                     do so.

(b) If two angles are both congruent to some third angle, then they are congruent to each other.

(c) If you have a line and you want to draw another line perpendicular to it, you may do so.

(d) If you have this situation

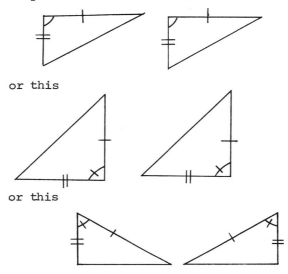

or this

or this

then the triangles would fit exactly on top of each other pair by pair, if for some reason you wanted them to. Also, the parts that weren't known to be congruent originally now can be marked congruent if you want to do so.

(e) If you have a segment and you want to draw another segment somewhere which is congruent to it, you may do so.

(f) If two segments are congruent to some third segment, then they are congruent to each other.

Axiom 8 is commonly called the "SAS" Axiom. Can you see why?

Exercise 4-22 Use Dedekind's Axiom, the Axiom of Separation, the Axioms of Connection, of Order, or of Congruence or any theorem or exercise proved so far in System 4 to answer the following questions:

(a) Explain why it's OK to draw a line connecting A to D if you want to.

27

(b) Explain why, if you begin extending the two endpoints of a line segment, they will never meet no matter how far you extend them.

(c) Explain why you may, if you want to, mark off a segment along XY with one endpoint at X and which is congruent to AB.

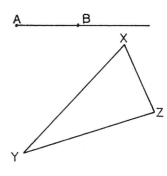

(d) Explain why you may, if you want to, draw a ray from C into the triangle, making an angle with CE which is congruent to angle RST.

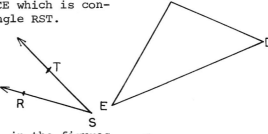

(e) Explain why, in the figures on the right, angle B is congruent to angle R, side BC is congruent to side RQ, and angle C is congruent to angle Q.

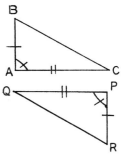

(f) Explain why a line is infinite in extent, that is, for any point on a line there is always another point on either side.

(g) Explain why angle DBC is congruent to angle BCA.

28

Here are abbreviations for some of the geometric concepts defined so far.

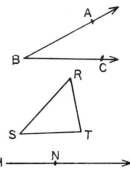

(1) For the word "angle" use < . So instead of writing "angle ABC," write < ABC. The symbol for "angles" is <S .

(2) For the word "triangle" use Δ . So instead of writing "triangle RST," write Δ RST. The symbol for "triangles" is 𝔄 .

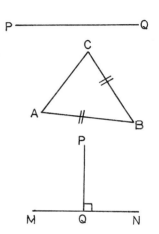

(3) For the word "ray" use the single-headed arrow ⟶ . So instead of writing "ray MN," write M͞N͢.

(4) For the word "segment" use a small segment — . So instead of writing "segment PQ," write P̄Q̄.

(5) For the word "congruent" or the phrase "is congruent to," use the symbol ≅ . So instead of writing "AB̄ is congruent to B̄C̄," write AB̄ ≅ B̄C̄.

(6) For the word "perpendicular" or the phrase "is perpendicular to," use the symbol ⊥ . So instead of writing "P̄Q̄ is perpendicular to M̄N̄," write P̄Q̄ ⊥ M̄N̄. The little block at Q is universally used in diagrams to indicate perpendicularity. It may be used on either side of PQ.

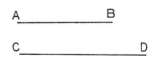

COMPARISON OF ANGLES AND SEGMENTS

It will be very useful in the work coming up to have a clearly defined meaning for what is meant when we say: (a) one segment is shorter (or longer) than another; (b) one angle is smaller (or larger) than another.

(a) If we have two segments AB̄ & C̄D̄ as shown, according to Axiom 1 of Congruence we can start at C and locate in the direction of D a point E which will make C̄Ē ≅ AB̄. If point E lands between C and D, then we can say "AB̄ is shorter than C̄D̄" and write AB̄ < C̄D̄, using the symbol " < " to mean "is shorter than."

(b) If we have two angles < ABC and < XYZ
     as shown, according to Axiom 5 of
     Congruence we could draw another ray
     with its endpoint at Y and make an
     angle with $\overrightarrow{YX}$ (or with $\overrightarrow{YZ}$) which is
     congruent to < ABC. If this ray is in
     the interior of < XYZ, then we can say
     " < ABC is smaller than < XYZ" and
     write < ABC $<$ < XYZ, using the
     symbol "$<$" to mean "is smaller than."

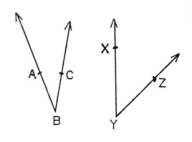

Exercise 4-23   (a) Write a definition to explain what is meant by
                     saying "$\overline{AB}$ is longer than $\overline{CD}$."

                (b) Write a definition to explain what is meant by
                     saying " < ABC is larger than < XYZ."

                (c) Explain why < ABC $<$ < ABD.

                (d) Explain why $\overline{MO} < \overline{MN}$.

AXIOMS OF TRICHOTOMY

1. For any two segments, the first is either shorter than, congruent
    to, or longer than the second.

2. For any two angles, the first is either smaller than, congruent to,
    or larger than the second.

ON KNOWING WHAT TO DO

     Theorems are statements that usually have the form of a simple
statement or an "If..., then..." statement. Proofs of theorems can be
direct or indirect.
     The biggest difficulty in doing a proof is knowing where to begin.
Three "Proof Rules" will tell you how to begin a proof under three
different circumstances. (Before beginning, review the meaning of
"hypothesis" and "conclusion" on page 13.)

Proof Rule 1 An indirect proof of a simple statement begins by sup-
              posing that the statement is false, that is, by denying
              the statement. Then try to find a contradiction to an
              axiom, a previously proved theorem, or a definition. An
              example of Proof Rule 1 is Theorem 4A, page 17.

30

<u>Proof Rule 2</u>   A direct proof of an "If..., then..." statement (an implication) usually begins with the hypothesis (the "If..." part) and uses axioms, previously proved theorems, or definitions to argue through to the conclusion (the "then..." part). An example of Proof Rule 2 is Theorem 4B, page 18.

<u>Proof Rule 3</u>   An indirect proof of an implication (an "If..., then..." statement) can begin with the hypothesis and then a supposition that the conclusion is false. Starting with this supposition, try to arrive at a contradiction. The proof of the following theorem is an example of Proof Rule 3.

Contrast Theorem 4C with Axiom 8 of Congruence.

<u>Theorem 4C</u>   If there are two triangles ABC and XYZ, and if $\overline{AB} \simeq \overline{XY}$, < B $\simeq$ < Y, and < A $\simeq$ < X, then $\triangle$ ABC $\simeq$ $\triangle$ XYZ.

(Alternate Form: If two angles and the included side of one triangle are congruent respectively to two angles and the included side of another triangle, then the triangles are congruent.)   The abbreviation is ASA.

<u>Proof</u>: Begin by drawing a sketch of the hypothesis, like this:

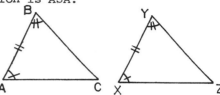

Now suppose that the conclusion is false, that is, that the triangles are <u>not</u> congruent. This means that $\overline{AC} \not\simeq \overline{XZ}$ (the symbol $\not\simeq$ means "is not congruent to") because if $\overline{AC}$ <u>were</u> congruent to $\overline{XZ}$, the triangles would be congruent by SAS.

Now if $\overline{AC} \not\simeq \overline{XZ}$, then either $\overline{AC} < \overline{XZ}$ or $\overline{AC} > \overline{XZ}$.   (Why?)

<u>Case 1</u>:   Suppose $\overline{AC} < \overline{XZ}$.

Then there is a point between X and Z (call it W) which will make $\overline{XW} \simeq \overline{AC}$, as shown.   (Why?)

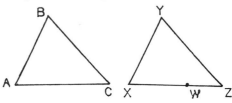

Now draw $\overline{YW}$, as shown.

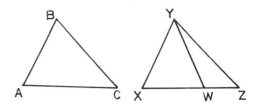

Thus $\triangle$ ABC $\simeq$ $\triangle$ XYW by SAS and so < B $\simeq$ < XYW. We already know that < B $\simeq$ < Y (< Y is < XYZ). Thus by Axioms 6 and 7 of Congruence, < XYW $\simeq$ < XYZ. Finish the proof.   (Look at Proof Rule 3 again.)

Learn the Proof Rules! They tell you what to do.

The next theorem is another "congruent triangles" theorem. We will not attempt to prove it; it can be proved, but the proof is rather complex. It is included here because it is useful. If you would like to see the proof, research it in another geometry book. Otherwise, just accept the fact that it can be proved.

Theorem 4D   If there are two triangles ABC and XYZ and if $\overline{AB} \simeq \overline{XY}$, $\overline{BC} \simeq \overline{YZ}$, and $\overline{AC} \simeq \overline{XZ}$, then the two triangles are congruent.

(Alternate Form: If three sides of one triangle are congruent respectively to three sides of another triangle, then the triangles are congruent.) The abbreviation is SSS.

So far there have been a "letter" axiom (SAS) and two "letter" theorems (ASA and SSS) on congruent triangles. There are other letter combinations, however, such as SSA or AAA or AAS which haven't been mentioned. These suggest other theorems on congruent triangles. As a further exercise, you might try deciding what such theorems would say and whether or not they would be true. (Of the last three letter combinations, only one suggests a true theorem.)

A subtle idea is used in the proof of Theorem 4C which many people know but do not realize they know. An example from algebra illustrates the idea, and a discussion follows.

Solve the equation: $x + 2 = 5$

Solution: To solve this equation in a mechanical way, subtract 2 from both sides of the equation, like this:

$$
\begin{array}{r}
x + 2 = 5 \\
-2 \quad -2 \\
\hline
x \quad\;\; = 3
\end{array}
$$

So 3 is the solution.

Now most people who know anything about algebra are aware that there is no _logical_ reason why 2 is subtracted from both sides. From a logical point of view, _any_ number could be subtracted from both sides (or any number added to, multiplied by, etc.) and the result would still be logically correct. However, only by subtracting 2 will you get the _solution_ to the equation. (Picture a slightly addled eccentric humming merrily to himself as he adds and subtracts the endless supply of numbers to both sides of the equation, content with merely being logically correct!) It is this idea which is behind meta-axiom 5.

Meta-Axiom 5   In the proof of a theorem, there may come a place in the argument where several steps could be correctly made, each in a different direction. In such a situation, take the step which leads to the conclusion you seek.

The question arises: When many things could be done, how does one decide what should be done? The answer: Do the thing that gets you where you want to go! The fun begins when you're not sure what that thing is.

Exercise 4-24   Find a step in the proof of Theorem 4C where the proof could logically have gone in a different direction.

Exercise 4-25   Discuss this paraphrase of meta-axiom 5: You may do whatever you like in a proof, as long as it is logically correct.

AXIOM OF PARALLELISM   (Playfair's Axiom)

If L is any line and if P is any point which is not on L, then there is at least one but no more than one other line in the same plane with L and P which passes through P and does not intersect L.

Make a sketch of what the Axiom of Parallelism says.

In the theorems that follow, some new ideas are mentioned, so before we get to the theorems, let's look at the ideas.

(1) In the figure on the right, lines
L$_1$ and L$_2$ are both intersected by line
L$_3$. When this happens, eight angles
are formed. The <u>interior angles</u> are
3, 4, 5, and 6. Angles 3 and 6 and
angles 4 and 5 are called <u>alternate-
interior pairs</u>. Name the alternate-
interior pairs formed by L$_1$, L$_2$,
and L$_4$.

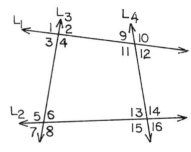
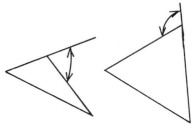

(2) <u>Exterior angle of a triangle</u>: For
the two triangles on the right,
exterior angles have been marked.
An exterior angle is formed by
extending one side of a triangle.
Then the angle between the extension and the <u>next</u> side is an
exterior angle. For each triangle
there are six (6) exterior angles.
Do you see why?

(3) <u>Straight angle</u>: A straight angle is an angle with two interiors.
If you have an angle and the sides "open" until they form a
straight line, the angle is called a straight angle.*

(4) <u>Vertical angles</u>: When two lines (it must
be two lines) intersect, four angles are
formed. In this situation two angles are
called vertical angles if they have a
common vertex but no common side. In the
sketch, angles 1 and 3 are one pair of
vertical angles. What is the other pair?

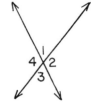

   (This use of the word "vertical" does not mean "up and down."
Instead it comes from the word "vertex." Vertical angles have the same
vertex.)

(5) <u>Supplementary angles</u>: If a ray is drawn
from some point on a line through a point
not on the line, two angles are formed,
as shown. Angles 1 and 2 together comprise a straight line and are called
supplementary angles.

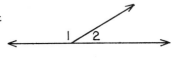

      Two angles do not have to share a
common side and vertex to be supplementary. For instance, the two angles
on the right might be supplementary.
This means that if one of the sides of

one of the angles is extended through the vertex to form a third
angle as shown, then that third angle is congruent to the other
angle.

---

* In other books, you may find a straight angle defined as an angle of
180 degrees. Degrees are not used in this book.

In the first figure on the right
<1 ≃ <3, so <1 is supple-
mentary to <2.

In the second figure, <2 ≃ <3,
so <2 is supplementary to <1.
(There is no symbolic abbreviation
for "is supplementary to.")*

(6) <u>Parallel Lines</u>: Two lines are parallel if they are in the same
plane but do not intersect.

Do not continue until you can make a sketch from memory of these
six things:

(1) alternate interior pairs of angles
(2) exterior angle of a triangle
(3) straight angle
(4) vertical angles
(5) supplementary angles
(6) parallel lines

Exercise 4-26   Explain why it is true that if an angle is greater than
a right angle, then an angle which is supplementary to it
must be less than a right angle. (Use an indirect proof
with the Axiom of Perpendicularity and the definition of
"smaller than" for angles.)

MORE ON DEFINITONS

Before reading further, take a few minutes to review pages 7 and 8.
Also review the meta-axioms.
We are going to take a closer look now at the role played by definitions in a deductive system.
Recall that definitions are used primarily to introduce new ideas
into a system. If one wishes to make and prove statements about a triangle or about parallel lines, these things must first be defined using
terms that are already in the system. A definition has all the logical
force of an axiom. (Some logicians consider a definition to be an
axiom.) The meta-axioms apply equally to definitions. Hence a definition should be paraphrased only with great care, lest meta-axiom 1 be
violated.
The process of defining new ideas or concepts using ideas or concepts that are already in the system has to stop eventually; that is,
there are some concepts or ideas that are never defined.

As an example, start with the definition of <u>perpendicular</u> (page 26).
(In the exposition that follows, you will learn a lot more if you look
back and read the definitions as they are referred to than if you don't.)

---

* The concept of "degrees" is not defined in this text. In many books
two angles are supplementary to each other if the sum of the angles is
180°, for example, angles of 120° and 60° are supplementary.

"Perpendicular" is defined using the concepts of "point," "line, "ray," "congruent," and "angle." Now point, line, and congruent have never been defined, but angle and ray are defined on page 22. An angle is defined using the concepts of "ray" and "line segment." A ray is defined using the concepts of "line" and "point" (concepts with no definitions). A (line) segment is defined (page 19) using the concepts "point," "line," and "between," none of which are defined.

Thus the definition of perpendicular, when traced back, depends upon four concepts that have never been defined: "point," "line," "congruent," and "between." Along with the concept "plane," these five comprise the undefined ideas or terms of the system. The definition of every other concept in System 4 can be traced back to them. They are called the "primitives" or "undefined terms" of the system.

The diagram below illustrates the dependence of the concept of "perpendicular lines" on the primitives. It should be looked at as a structure sticking into the air with the five primitives on the ground.

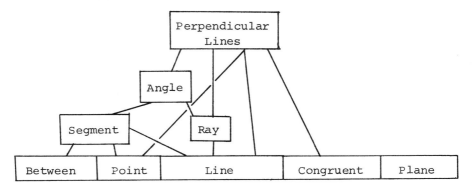

The fact that the primitives have no definition is not an oversight. It is intentional. If one attempts to define them, other concepts must be used in the definition. Then these other concepts must themselves be defined using still other concepts which must be defined, etc. This is a never-ending process, so we <u>intentionally</u> stop it somewhere and say "Here are the primitives and this is where we begin."

The primitives of a system should always be stated at the outset, even before the axioms. This was not done with System 4 because we were concentrating on other things. It will be done from now on, however.

To say that a concept has no definition is not the same as saying that it has no meaning. Instead, since it is not defined, it can mean anything you want it to as long as the axioms and meta-axioms are not violated.

For instance, most of us learn as children the meaning of the concept of a line. However, when "line" is explicitly given as a primitive concept of a system, you may use the meaning you learned as a child if it's convenient to do so. But, if the axioms or meta-axioms require it (as they will in System 5), be prepared to change the meaning or mental picture you have of a line.

Exercise 4-27   In the diagram below, draw connecting lines to show how
the definition of <u>supplementary angles</u> depends on the
primitives.

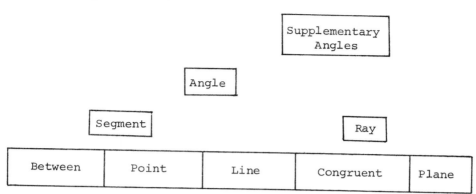

Exercise 4-28   Draw a diagram showing how the definition of <u>exterior
angle of a triangle</u> depends on the primitives of the
system.

Introducing new concepts into a system by devising ingenious and
properly worded definitions is one of the ways of expressing mathematical
creativity.

Exercise 4-29   Criticize the following definitions:
        (a)   Parallel lines are lines which don't intersect.
        (b)   Supplementary angles are angles which add up to 180°.
        (c)   Vertical angles are two angles which have the same
              vertex.
        (d)   An angle is two lines coming together.
        (e)   AB is shorter than CD if it doesn't have as much
              length.

Exercise 4-30   How could the concept of "angle measure" (degrees) be
introduced into System 4?   (If you introduce this concept
properly, you may use it from now on if you so desire.)

For quick reference, the axioms of Euclidean Geometry (System 4) are
reassembled here.

Axioms of Connection (page 16)

1. For any two distinct points, there is exactly one line through them both.

2. There are at least two points on every line.

3. There are at least three points on every plane, not all of which are on the same line.

4. For any (every) three distinct points that do not lie on the same line, there is exactly one plane on all three.

5. If two points of a line lie on a plane, then the entire line lies on that same plane.

6. There are at least four points, not all of which lie on the same plane.

7. If two planes have one point in common, then they have exactly one line in common.

Axioms of Order (page 19)

1. For any two points of a line, there exists at least one third point of the line which lies between them.

2. For any three points on a line, exactly one lies between the other two.

3. Every segment has exactly one midpoint.

4. If a line and a triangle are in the same plane and if the line intersects a side of the triangle at a point which is not a vertex, then the line also intersects one of the other sides of the triangle.

Dedekind's Axiom (page 20)

If all points on a line are separated into two nonempty sets L and R in such a way that every point of L is to the left of every point in R, then there is exactly one point T which is the boundary of this separation and T is either the rightmost point of L or the leftmost point of R.

Axiom of Separation (page 21)

Every line partitions every plane in which it lies into two half-planes called the left half-plane and the right half-plane (or the upper and lower half-planes) in such a way that:

(1) Exactly one of the following is true: every point of the plane is in the left half-plane or in the right half-plane or on the line.

(2) For any point in one half-plane and any point in the other, the line joining them intersects that line which formed the half-planes.

(3) If two points of a line are in the same half-plane, then every point of the line between them is in that half-plane.

Axioms of Congruence (pages 25 and 26)

1. If AB is a segment of a line L and C is a third point on the same or some other line, then there is on either side of C and on the same line exactly one fourth point D which will make AB congruent to CD.

2. Every segment is congruent to itself. If any segment is congruent to a second segment then the second segment is congruent to the first.

3. If seg AB is congruent to seg CD and seg CD is congruent to seg EF, then seg AB is congruent to seg EF.

4. Axiom of Perpendicularity   If L is any line and P is any point, then there is exactly one line through P which is perpendicular to L.

5. Suppose ABC is any angle. If O is a point on some line L, and M is a second point on L, then (on either side of L) there is exactly one ray with its endpoint at O which forms an angle with ray OM which is congruent to angle ABC.

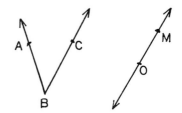

6. Every angle is congruent to itself. If any angle is congruent to a second angle then the second angle is congruent to the first.

7. If angle ABC is congruent to angle POR and angle POR is congruent to angle XYZ, then angle ABC is congruent to angle XYZ.

8. If there are two triangles (call them ABC and XYZ) and if AB is congruent to XY and angle ABC is congruent to angle XYZ and BC is congruent to YZ, then triangle ABC is congruent to triangle XYZ. (Also stated: If two sides and the included angle of one triangle are congruent respectively to two sides and the included angle of another, the triangles are congruent.)

Axioms of Trichotomy (page 30)

1. For any two segments, the first is either shorter than, congruent to, or longer than the second.

2. For any two angles, the first is either smaller than, congruent to, or larger than the second.

Axiom of Parallelism (Playfair's Axiom)   (page 33)

If L is any line and if P is any point which is not on L, then there is at least one but no more than one other line in the same plane with L and P which passes through P and does not intersect L.

Theorem 4E   If two angles are each supplementary to some third angle, then the two angles are congruent.

Proof: (This is a direct proof, so look at Proof Rule 2, page 31, before beginning.)

The proof begins by drawing a picture
of the hypothesis, as shown at the
right, with<\S 1 and 2 both supplementary
to < 3.

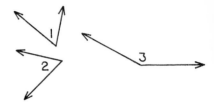

Now according to definition (5) on
page 34 for supplementary angles, if
a side of < 3 is extended through the
vertex (forming < 4) as shown, then
both < 1 and < 2 are congruent to < 4.
In abbreviated language we can write
< 1 ≃ < 4 and < 2 ≃ < 4. Now use
Axioms 6 and 7 of Congruence to
finish the proof.

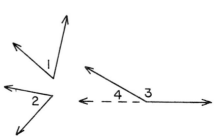

Exercise 4-31   If we have four angles (Call them angles 1, 2, 3, and 4)
and if < 1 is supplementary to < 2 and < 2 ≃ < 3, and
if < 3 is supplementary to < 4, how are<\S 1 and 4 related
and why?

Now use Theorem 4E to prove the following theorem.

Theorem 4F   If two lines intersect, the vertical angles so formed are
congruent.

Proof: This is also a direct proof. Draw a picture of the hypothesis
and then use the definition of supplementary angles and Theorem 4E to
reach the conclusion.

Study hint: Knowing (or memorizing) a proof is no guarantee that
you will understand it. However, understanding begins with knowing.
It is certain that if you don't know the proof, you won't understand it.
Try learning proofs even when you don't fully understand them.   Then
your mind has something to work on.

<u>Theorem 4G</u>   An exterior angle of a triangle is greater than either of
           the two interior angles that are not adjacent to it.

<u>Proof:</u>   In the figure on the right, we must
show that < 4 > < 1 and < 4 > < 2.

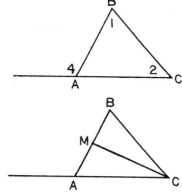

Start with proving < 4 > < 1.   Pick the
<u>midpoint</u> of $\overline{AB}$, call it M, and draw a
segment from C to M as shown (Axiom 1 of
Connection).

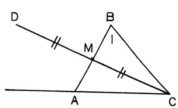

Now extend $\overline{CM}$ through M to form another
segment congruent to $\overline{CM}$ as shown, call
the endpoint D (Axiom 1 of Congruence).

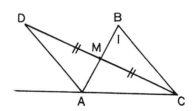

Now connect D to A (Axiom 1 of
Connection).

It is now possible to make two further deductions: $\overline{AM} \cong \overline{MB}$.   Why?

< DMA $\cong$ < BMC.   Why?

Therefore $\triangle$ DMA $\cong$ $\triangle$ BMC.   Why?

And so < MAD $\cong$ < 1.   Why?

Now finish the proof (see exercise 4-23(c), page 30).

Exercise 4-32   Use Theorem 4G to explain why a triangle cannot have
                more than one right angle or more than one angle which
                is greater than a right angle.

Theorem 4H   If L_1 and L_2 are any two lines in the same plane and if they
             are both intersected by some third line and if the alternate
             interior angles so formed are congruent, then L_1 and L_2 do
             not intersect.

Proof: The proof is indirect (see Proof Rule 3 page 31).   Begin by
making a sketch of the hypothesis and use Theorem 4G.

Exercise 4-33   Suppose two lines are intersected by a third line in
                such a way that the alternate interior angles are not
                congruent.   Can you infer from Theorem 4H that the lines
                do not intersect?

     Students having difficulty should redo exercises 4-21 and 4-22,
pages 26 and 27, before continuing.

     The following theorem is the logical converse of Theorem 4H.   The
converse of an implication is another implication formed from the first
implication by switching the hypothesis (the "If..." part) and the conclusion (the "then..." part).   The truth of an implication does not
imply the truth of its converse.   (Sometimes the converse of an implication is true, sometimes not.)   The converse must be proved as a
separate theorem.

Theorem 4J   If L_1 and L_2 are any pair of parallel lines (notice this is
             not the same way Theorem 4H begins!) and if they are both
             intersected by some third line, then the alternate interior
             angles that are formed must be congruent.

Proof: This proof is also indirect.   Begin the usual way by making a
sketch of the hypothesis.   Somewhere in your proof you must use Axiom 5
of Congruence, the Axiom of Parallelism, and Theorem 4H.

Exercise 4-34   Use Theorem 4J to show how it is possible to have two
                triangles whose angles are congruent but whose sides are
                not.   (Such triangles are called "similar," see Appendix
                A.)   Begin with a triangle and draw a line intersecting
                two sides and parallel to the third.

Definition: Two angles are said to comprise
a third angle if a ray can be drawn with its
endpoint at the vertex of the third angle
and forming with the sides of the third
angle two angles which are congruent
respectively to the first two angles.   To
see this illustrated, note that in the
figure on the right, angles 1 and 2 comprise angle ABC because a ray can be
drawn from the vertex of angle ABC
forming angles which are congruent to
angles 1 and 2.

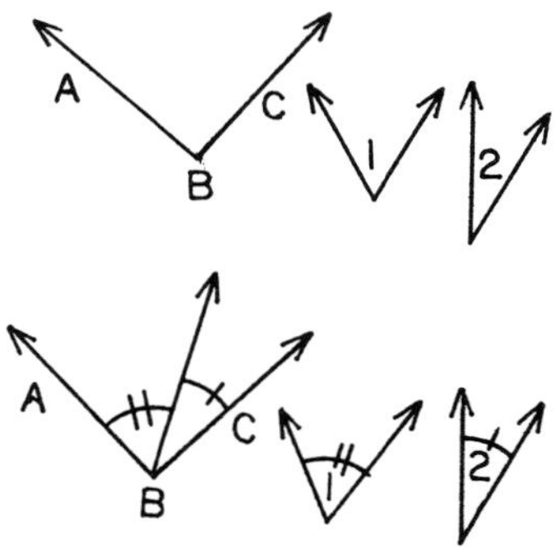

Exercise 4-35   Devise a definition for the phrase: Three angles
                comprise a fourth angle.

Theorem 4K   The angles of any triangle comprise one straight angle.

Proof: The proof uses the Axiom of Parallelism and Theorem 4J. Whether direct or indirect is one of the things you have to discover. Begin by drawing a triangle.

Exercise 4-36   Explain how Theorem 4E was used in the proof of Theorem 4K.

This is as far as this book takes Euclidean geometry. There are, however, hundreds of theorems in this system, which shouldn't be surprising, since men have been adding theorems to the system for over 2,000 years. Some of the theorems pre-date Euclid himself.

There are at least three kinds of mathematically creative activity:
(1)   You read or hear a statement and then devise a proof for it, thus making it a theorem;
(2)   You think of a statement yourself (called a "conjecture") and try to devise a proof for it;
(3)   You devise your own axioms (using the meta-axioms as a guide) and definitions and build a system of your own.
Theoretical mathematicians engage in all of these activities.

If you find one of these activities diverting or amusing, it needs no further defense. If you have learned a little to enjoy "doing proofs," then you have expanded the number of activities in this world which you can enjoy. This is a true growth experience. One can ask for little more from math.

If you'd like to pursue Euclidean geometry a little further on your own, some statements are provided below. See if you can prove them.

(1)   Any exterior angle of a triangle is comprised by the two non-adjacent interior angles. (Compare this theorem to Theorem 4G.)

Definitions: If two sides of a triangle are congruent, the triangle is called isosceles. If all three sides are congruent, the triangle is called equilateral.

Isosceles Δ   Equilateral Δ

(2)   In an isosceles triangle, the angles opposite the congruent sides are congruent. (Hint: Use an indirect proof and ASA.)

(3)   Each of the angles of an equilateral triangle is congruent to each of the others. (Hint: Use Theorem 2 above and Axiom 7 of Congruence.)

<u>Definitions:</u>   The noncongruent side of an
                     isosceles triangle is called
                     the <u>base</u>. The angle
                     opposite the base is called
                     the <u>vertex angle</u>.

                     A ray or segment is said to
                     <u>bisect</u> an angle if it passes
                     through the vertex and
                     interior of the angle and
                     forms two congruent angles
                     with the sides.
                     (See figure.)

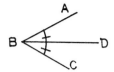

$\overline{BD}$ bisects < ABC

                     A line or segment is said
                     to <u>bisect</u> a segment if it
                     passes through the mid-
                     point of the segment.
                     (See figure.)

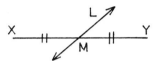

Line L bisects XY

(4) If a line bisects the vertex angle of an isosceles triangle, then
     it bisects the base and is perpendicular to the base. (Hint: Use
     the definition of perpendicular lines, page 26.)

(5) If a line bisects the base of an isosceles triangle and is perpen-
     dicular to the base, then it passes through the vertex of the
     vertex angle. (Use an indirect proof. Hint: Use Axiom 1 of
     Connection and Axiom 4 of Congruence.)

     Statement 5 should suggest another statement to you. If it does,
     state it and prove it.

(6) If two angles and a non-included side of one triangle are congruent
     respectively to two angles and a non-included side of another, then
     the triangles are congruent. This theorem is labeled AAS. (The
     proof is similar to that for ASA.)

(7) If L_1 and L_2 are two parallel lines and if a segment is drawn with
     one endpoint on L_1 and one on L_2 and if M is the midpoint of that
     segment and if another segment is drawn through M with its end-
     points on L_1 and L_2 respectively, then M is also the midpoint of
     the second segment. (Begin by drawing a picture which includes
     all of the "If..." parts.)

(8) If a line is perpendicular to one of two parallel lines, then it
     is perpendicular to the other. (Hint: Use the definition of
     perpendicular lines on page 26 and Theorem 4J.)

(9) If L_1 and L_2 are parallel lines, and if $\overline{AB}$ and $\overline{CD}$ are segments
which are parallel to each other with points A and C on L_1 and
points B and D on L_2, then $\overline{AB} \cong \overline{CD}$. (Begin by making a sketch of
the hypothesis, then use Theorem 4C.) Note that the Theorem does
not say that AB and CD must be perpendicular to L_1 and L_2.

Definitions: If $\overline{AB}$, $\overline{BC}$, $\overline{CD}$, and $\overline{DA}$ are four
segments, and $\overline{AB}$ and $\overline{CD}$ have no
points in common, and $\overline{BC}$ and $\overline{DA}$
have no points in common, then
the four segments comprise a
figure called a quadrilateral
(abbreviated 4-lat). See Fig. 1.
(It is considered undesirable to
have a figure like Fig. 2 called
a 4-lat, that is why we say "$\overline{BC}$
and $\overline{DA}$ have no points in common,"
etc.)

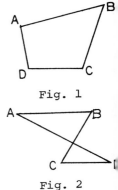

Fig. 1

Fig. 2

The four segments are called the sides of the 4-lat and the four endpoints are its vertices. The angles of the 4-lat are the angles formed
by a pair of adjacent segments. Since there are 4 pairs of adjacent
segments, the 4-lat has four angles.

Angles of the 4-lat which do not share a common side are called opposit
angles. In Fig. 1,<S A and C are opposite<S , as are<S B and D.

Sides of the 4-lat that do not share a common vertex are called opposit
sides. In Fig. 1, $\overline{AB}$ and $\overline{CD}$ are opposite sides, as are $\overline{AD}$ and $\overline{BC}$. Sid
that do share a common vertex are called adjacent.

A parallelogram is a 4-lat for which both pairs of opposite sides are
parallel. (If only one pair of opposite sides are parallel, the 4-lat
is called a trapezoid.)

(10) If one angle of a parallelogram is a right angle, then so are the
other three angles.

(11) Opposite sides of a parallelogram are congruent. (Put in the
"if..., then..." form, this statement would be: If a figure is a
parallelogram, then its opposite sides are congruent.) See
Theorem 9 above.

(12) Opposite angles of a parallelogram are congruent.

More topics in Euclidean geometry (similar triangles and the
Pythagorean Theorem) can be found in the appendices.

ON CHANGING AXIOM SETS

We're about to change axioms again, using the meta-axioms as a guide. Review meta-axiom 1 (page 8), meta-axiom 2 (page 11), meta-axiom 3 (page 16) and meta-axiom 4 (page 18) before continuing.

Since axioms are only assumptions, students often wonder why they can't simply assume anything they like. The answer is that you <u>can</u>, within limits. These limits are stated by the meta-axioms.

<u>Meta-Axiom 6</u> (Consistency) No set of axioms may contain two statements which contradict each other, nor may it contain statements which lead to contradictory theorems.

Exercise 4-37 Explain why the following collection of statements could not serve as axiom set for a deductive system:
1. Horses, chickens, dogs, pigs, and ducks are animals.
2. All animals are created equal.
3. There is at least one animal of every kind.
4. If an animal has four legs, a snout and goes "oink-oink," then it is superior to other animals.
5. Pigs have four legs, a snout, and go "oink-oink."

Meta-axiom 4 (page 18) says in effect that you have to have enough statements in your axiom set to enable you to "get somewhere." Hence you mustn't have too few statements. It might be, however, that when trying to choose a set of axioms, you have too many statements. The following meta-axiom covers that possibility.

<u>Meta-Axiom 7</u> (Independence) No axiom may be provable from the other axioms of the set.

Of course, if an axiom is provable from some of the others, it could be a theorem. This criterion of independence of axioms is often violated in introductory courses like this one. The reason is expediency. In an introductory course, one must try to strike a balance between meeting meta-axiom 7 and bewildering students with complex and subtle logical arguments.

TRANSITION

The following exercise is important in carrying out the transition from System 4 (Euclidean geometry) to System 5. You should spend some time trying to answer it correctly.

Exercise 4-38 Which of the theorems that we have had in Euclidean
geometry would have to dropped from the system if the
Axiom of Parallelism were taken out of the axiom set?

Exercise 4-39 Is it possible to prove from our everyday experience that
the Axiom of Parallelism must be true in the world around
us?

At the end of System 5 (pages 69 and 70) is a list of readings in
non-Euclidean geometry. You may wish to do those readings before beginning System 5 (Lobachevskian geometry), though in the author's
opinion, they will mean more to you after you have studied System 5.

System 5

lobachevskian geometry

The terms "point," "line," "plane," "congruent," and "between" are primitive.

AXIOMS OF CONNECTION

1. For any two distinct points, there is exactly one line through them both.

2. There are at least two points on every line.

3. There are at least three points on every plane, not all of which are on the same line.

4. For any (every) three distinct points that do not lie on the same line, there is exactly one plane on all three.

5. If two points of a line lie on a plane, then the entire line lies on that same plane.

6. There are at least four points, not all of which lie on the same plane.

7. If two planes have one point in common, then they have exactly one line in common.

AXIOMS OF ORDER

1. For any two points of a line, there exists at least one third point of the line which lies between them.

2. For any three points on a line, exactly one lies between the other two.

3. Every segment has exactly one midpoint.

4. If a line and a triangle are in the same plane and if the line intersects a side of the triangle at a point which is not a vertex, then the line also intersects one of the other sides of the triangle.

DEDEKIND'S AXIOM

If all points on a line are separated into two nonempty sets L and R in
such a way that every point of L is to the left of every point in R,
then there is exactly one point T which is the boundary of this separation and T is either the rightmost point of L or the leftmost point of F

AXIOM OF SEPARATION

Every line partitions every plane in which it lies into two half-planes
called the left half-plane and the right half-plane (or the upper and th
lower half-planes) in such a way that:
(1) Exactly one of the following is true; every point of the plane is in
    the left half-plane or in the right half-plane or on the line.
(2) For any point in one half-plane and any point in the other, the line
    joining them intersects the line which formed the half-planes.
(3) If two points of a line are in the same half-plane, then every point
    of the line between them is in that half-plane.

AXIOMS OF CONGRUENCE

1. If AB is a segment of a line L and C is a third point on the same
    or some other line, then there is on either side of C and on the
    same line exactly one fourth point D which will make AB congruent
    to CD.

2. Every segment is congruent to itself. If any segment is congruent
    to a second segment, then the second segment is also congruent to
    the first.

3. If seg AB is congruent to seg CD and seg CD is congruent to seg EF,
    then seg AB is congruent to seg EF.

4. Axiom of Perpendicularity   If L is any line and P is any point,
    then there is exactly one line through P which is perpendicular to

5. Suppose ABC is any angle. If O is
    a point on some line L and M is a
    second point on L, then (on either
    side of L) there is exactly one ray
    with its endpoint at O which forms
    an angle with ray OM which is con-
    gruent to angle ABC.

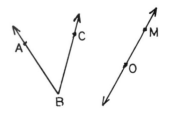

6. Every angle is congruent to itself. If any angle is congruent to
    a second angle, then the second angle is also congruent to the firs

7. If angle ABC is congruent to angle POR and angle POR is congruent
    to angle XYZ, then angle ABC is congruent to angle XYZ.

8. If there are two triangles ABC and XYZ and if AB is congruent to XY
    and angle ABC is congruent to angle XYZ and BC is congruent to YZ,
    then triangle ABC is congruent to triangle XYZ.

    (Also stated: If two sides and the included angle of one triangle
    are congruent respectively to two sides and the included angle of
    another, the triangles are congruent.)

AXIOMS OF TRICHOTOMY

1. For any two segments, the first is either shorter than, congruent to,
    or longer than the second.

2. For any two angles, the first is either smaller than, congruent to,
    or larger than the second.

AXIOM OF NON-INTERSECTION

If L is any line and if P is any point which is not on L, then there are
at least two other lines in the same plane with L and P which pass
through P and do not intersect L.

    All of the definitions from System 4 (page 19, pages 22 through 25,
page 29, page 30, and pages 34 and 35) carry over to System 5 with a
single exception: the definition of parallel lines on page 36. Parallel
lines are no longer "two lines in the same plane which do not intersect."
A new definition will be given shortly.

PRELIMINARIES

Exercise 5-1   Take a few minutes to contrast the axiom sets of System 4
               (Euclidean geometry) and System 5 (Lobachevskian geometry),
               that is, tell how they are different and how they are the
               same. (Don't continue until you have done this.)

    As you have probably noticed, most of the axioms of Euclidean
geometry are also axioms of Lobachevskian geometry. This means that if
a theorem was proved in System 4 using only axioms and definitions which
carry over into System 5, then the theorem carries over also.
    For example, the proof of Theorem 4C (page 31, the ASA Theorem)
required the use of the definitions of "greater than" for segments and
for angles, and also Axiom 1 of Connection, and Axioms 6, 7, and 8 of
Congruence. (Verify this for yourself.) Since all of these definitions
and axioms are also part of System 5, Theorem 4C (the ASA Theorem) is
also true in System 5.

Exercise 5-2   Explain why Theorem 4H is also true in System 5 but
               Theorem 4J may not be true.

Exercise 5-3   List the theorems of System 4 which are also theorems in
               System 5 and those which are not.   (Spend at least 20
               minutes on this exercise before continuing.)

LAYING THE GROUNDWORK

Lemma If a line L intersects a triangle at
       a vertex (and if the line and triangle
       are in the same plane)   and if part of
       the line lies in the interior of the
       triangle (as shown), then L intersects
       the opposite side.

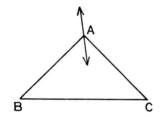

Exercise 5-4   To prove the lemma extend $\overline{AC}$ through A any length.   Call
               the endpoint D.   Join D to B.   Then use $\triangle BCD$, Axiom 4 of
               Order, and the Axiom of Separation to explain why the con-
               clusion must be true.

       You may have found the Axiom of Non-Intersection a surprise.   (It
is talking about straight lines!   Look again at meta-axiom 1, page 8.)
You may find the following theorem even more surprising.

Theorem 5A   For any line L and a point P which is not on L, there are
             an infinite number of lines in the plane with L and P which
             pass through P which do not intersect L.

Proof: Start with a Line and a point P not on L (Proof Rule 2, page 31).

According to the Axiom of Non-Intersection there are at least two lines
through P which do not intersect L.   Draw them in and call them L_1 and
L_2.

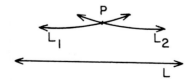

Now choose point A on L_1 and point B on L_2, draw the line through them, and call it L_3, as shown.

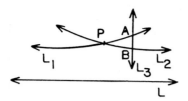

Now pick a point on L_3 between A and B (call it X) and draw the line through P and X (See exercise 4-18, page 24). Use an indirect proof to explain why the line through P and X does not intersect L. (Hint: Draw a line through P perpendicular to L and use the lemma.)

Since any point between A and B can be joined to P by a line which does not intersect L, there are an infinite number of lines through P which do not intersect L, which is what we wanted to prove.

Exercise 5-5 Suppose you pick a point on L_3 which is not between A and B (and is neither A nor B) and join that point to P with a line. Can you conclude that this line does (or does not) intersect L?

We know that if L is any line and P is any point not on L, we can draw exactly one line through P which is perpendicular to L, as shown below. (If this perpendicular is not always shown in the drawings that follow, you can always put it in yourself.)

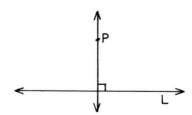

The <u>right</u> <u>side</u> or "on the right" refers to the right half-plane formed by this perpendicular. The <u>left</u> <u>side</u> or "on the left" refers to the left half-plane formed by this perpendicular.

There are an infinite number of lines through P that do not intersect L. There are also an infinite number of lines through P that <u>do</u> intersect L. (Hint: Pick points on L.) Do not continue without trying to figure out why these two statements are true.

Now to develop these ideas a little further:

In the figure on the right, a number
of lines have been drawn, all of which
pass through P and are in the plane with
L. Some of these lines intersect L and
some do not. Thus we get two sets of
lines: the set of all lines through P
that do not intersect L and the set of
all lines through P that do intersect L.

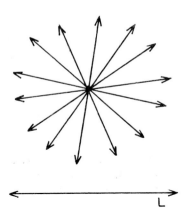

Can these two sets of lines overlap? That is, is it possible
moving clockwise around P to have a few lines through P that do not
intersect L, and then a few that do, and then a few more that do not,
etc.?

Try to answer this question before continuing. Use the space
below for sketches.

There is no overlapping of the two types of lines. Don't continue
until you understand why this is so.

A NEW MEANING FOR "PARALLEL"

In order to introduce a new
meaning for the concept of parallelism,
consider line L, point P, and a second
line L_1 perpendicular to L but not

passing through P, as shown on the
right. (In what follows, imagine a
line through P perpendicular to L,
even though it's not shown.)

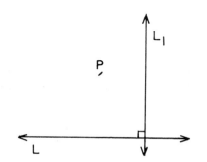

If each point of L_1 is joined to P with a line, then the points of L_1 are divided into two classes: those points which, when joined to P, determine a line that intersects L on the right and secondly, those points of L_1 which, when joined to P, determine a line that does not intersect L on the right. (See figure on the right. Look again at the definition of <u>right side</u> on page 53.)

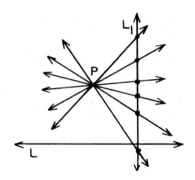

Since there is no overlapping of intersecting and non-intersecting lines, the points of L_1 are divided into two sets, which do not overlap and which together make up all of L_1. Now use Dedekind's Axiom (page 50) to explain why the following statement is true:

There is a boundary line which separates the lines (through P) that intersect L from the lines that do not.

WHICH SET OF LINES DOES THE BOUNDARY LINE BELONG TO? That is, does the boundary line (which separates the intersecting lines from the non-intersecting lines) intersect L?

There are only two possibilities:

(1)   If the boundary line intersects L, then starting with the perpendicular to L through P and moving counterclockwise around P, it is the last line to do so. Don't continue until you see that this is so.

(2)   If the boundary line does not intersect L, then starting with the perpendicular to L through P and moving counterclockwise around P, it is the <u>first one</u> not to intersect L. Don't continue until you see that this is so.

The entire argument above could be repeated if L_1 were drawn on the left side.

<u>Theorem 5B</u>   There exists a first non-intersecting line through P on the right and one on the left.

Theorem 5B says that the second possibility above is correct. Prove this below, using an indirect proof. (Suppose that the first possibility is true and arrive at a contradiction. See Proof Rule 1, page 30.)

Definition of "Parallel Through a Point":
Of all the lines through P which do not
intersect L, the <u>first one</u> on the right
is called the <u>right-hand parallel to L</u>
<u>through P</u>. The <u>first one</u> on the left is
called the <u>left-hand parallel to L</u>
<u>through P</u>. The other non-intersecting
lines are just called "non-intersecting."

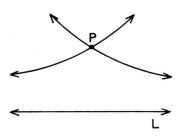

Memorize this definition before continuing.

Exercise 5-6   Is the right-hand parallel to L through P also parallel to
               L <u>on the left</u> or merely non-intersecting?

Exercise 5-7   If L is any line and P is any point not on L, is it always
               possible to draw a line through P which is right-hand
               parallel to L?

Two lines L_1 and L_2 (shown on the
right) are said to be <u>right-hand parallel</u>
if, when a line is drawn from a point P on
L_2 perpendicular to L_1, then L_2 is right-
hand parallel to L_1 through P, that is,
L_2 is the first non-intersecting line
through P on the right. A similar
meaning is given for left-hand parallel
lines.

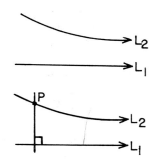

Exercise 5-8   Can two lines in the same plane not intersect and still
               not be parallel?

INFINITE TRIANGLES AND ANGLES OF PARALLELISM

<u>Definition of an Infinite Triangle</u>: If we have two lines which are
right (or left)-hand parallel and we connect a point on one to a point
on the other by a line segment, the resulting figure is called an
infinite triangle. (Note that an infinite triangle only has two angles.)

Examples of Infinite Triangles:

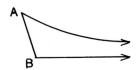

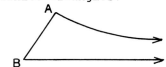

     All the lines drawn in this system should be considered <u>straight</u>
lines. We do not deal with any other kind of line. If the sketches do
not agree with your notion of straightness, keep in mind that they don't
have to! The sketches must agree with the axioms and theorems, not with
our preconceived notions of "straightness." See meta-axiom 1, page 8.

     If two angles or two segments are congruent, we have a way of
showing this fact by using equal numbers of slashes on the segments or
angles.

For example, in the figures on the right,
congruent segments and congruent angles
have been marked.

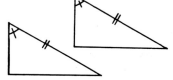

     Until now, we have not had a way of marking two lines parallel.
We could <u>say</u> they were parallel, but we had no way of marking them as
such. Now there's a way!

If we wish to indicate in a sketch that two
lines in the same plane are right- or left-
hand parallel (this is not the same as
saying that they do not intersect), we will
write the Greek letter omega, Ω, between
them, as shown on the right.

If a point on one of the lines is joined
to a point on the other, as shown, and
the points are labeled A and B, the
infinite triangle so formed will be
called $\triangle AB\Omega$.

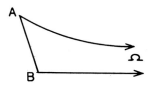

<u>Lemma</u>:   If L_1 and L_2 are right-hand parallel, as shown,

and if a segment is drawn from a point on L_1 to L_2, as shown,

and if the triangle is labeled as shown,

and if another line L_3 is drawn through A that makes an angle with $\overline{AB}$ which is smaller than $< BA\Omega$, as shown,

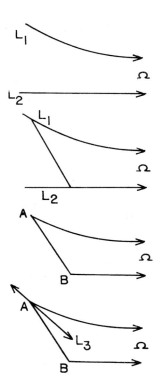

<u>then</u> L_3 must intersect ray $B\Omega$, that is, L_3 must intersect L_2.

To prove this, draw a segment from A perpendicular to L_2 and use the definition of "right-hand parallel through a point" on page 56, i.e., $\overrightarrow{A\Omega}$ (L_1) is the <u>first</u> non-intersecting line through A.

<u>Definition of Obtuse and Acute Angles</u>: An angle is called <u>acute</u> if it is smaller than a right angle. An angle is called <u>obtuse</u> if it is greater than a right angle but less than a straight angle.

   In the figure, $< COD$ is acute and $< AOD$ is obtuse.

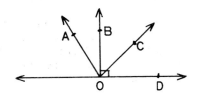

Definition of Angle of Parallelism:
If L is any line and P is any point not
on L, as shown,

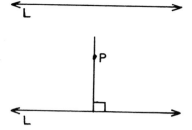

and if a line is drawn through P which is
perpendicular to L, as shown,

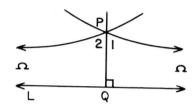

then the angle between this perpendicular
and the right-hand parallel (angle 1 in the
figure) is called the right-hand angle of
parallelism.

A similar definition holds for the left-hand
angle of parallelism.

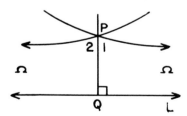

The angle of parallelism is the angle between the parallel and the
perpendicular.

Exercise 5-9   Explain why, in the sketch above, the right-hand parallel
               line could not be the same line as the left-hand parallel.

Theorem 5C   The left-hand angle of parallelism is congruent to the
             right-hand angle of parallelism.

Proof: Begin with the sketch on the right.

The theorem is a simple statement and the proof is indirect, so look
at Proof Rule 1, page 30, before continuing.

We wish to prove that $<1 \cong <2$, so begin by supposing that this isn't
so, that is, $<1 \not\cong <2$.

Then either $<1 > <2$ or $<1 < <2$.   Let's begin by supposing that
$<1 > <2$.

Then draw another line (call it L_4) through P which makes an angle with $\overline{PQ}$ congruent to < 2 (Axiom 5 of Congruence) as shown on the right.

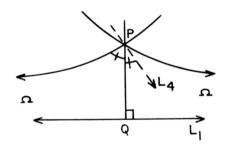

Now apply the previous lemma to draw a conclusion before continuing.

Call the point where L_4 intersects L_1 R, as shown. Then mark off on the left side of Q a segment which is congruent to $\overline{QR}$. Call its endpoint S as shown, so $\overline{SQ} \cong \overline{QR}$.

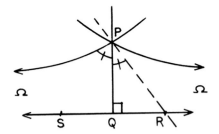

Now connect P to S as shown. Now △ PQR ≅ △ PQS (why?) and so < SPQ ≅ < QPR.

But line L_4 was drawn so that < QPR ≅ < 2.

Finish the proof.

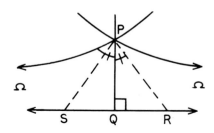

Theorem 5D The right and left-hand angles of parallelism are acute.

Proof: Suppose they were not acute. Then they would either be right angles (make a sketch) or obtuse angles (make a sketch). Both situations are impossible. Explain why. (Use Axiom 4 of Congruence for the first part and the definitions of "angle of parallelism" and "parallel through a point" for the second.)

60

Exercise 5-10   Explain why it is true that if a line is perpendicular to
                one of two parallel lines, it is not perpendicular to the
                other, but instead must make an acute angle with it.
                (Make a sketch.)

<u>Theorem 5E</u>   The exterior angle of an infinite triangle, formed by
                extending $\overline{AB}$, is greater than the opposite interior angle.

<u>Proof:</u>   Begin by sketching an infinite
triangle as shown. Call it △ ABΩ .
Then extend $\overline{AB}$ and mark angles 1, 2,
and 3, as shown.

We must show that < 3 > < 1.   The
proof is indirect.

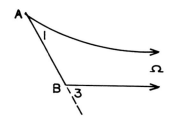

<u>Supposition 1:</u>   < 3 ≅ < 1. From M,
the midpoint of $\overline{AB}$, draw a perpendicular to the extension of $\overrightarrow{B\Omega}$, as
shown (Axiom 4 of Congruence). Call
the point of intersection C.

From A, mark off a segment on $\overrightarrow{A\Omega}$
that is congruent to $\overline{CB}$. Call its
endpoint D as shown.

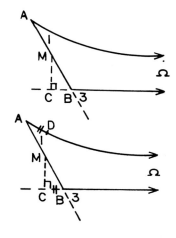

Now connect D to M. (Notice that we cannot tell whether $\overline{DM}$ and $\overline{MC}$ are part of the same line or not, even though they may look like it.)

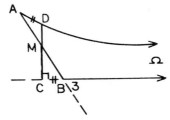

Since we are supposing that $<3 \cong <1$, it must be true that $<MBC \cong <1$. Thus $\triangle ADM \cong \triangle CBM$ (why?).

This means that $<AMD \cong <CMB$ and since $\overline{AB}$ is a single line, $\overline{DM}$ and $\overline{MC}$ must be part of the same line (suppose not!). Furthermore, $<ADM$ must be a right angle. (Why?) Thus $\overline{DC}$ is a single segment which is perpendicular to both $\overrightarrow{A\Omega}$ and $\overrightarrow{B\Omega}$.

Now find the contradiction and finish this part of the proof. (Hint: Look at exercise 5-10.)

Supposition 2: $<3 < <1$. Draw a line through A that makes an angle with $\overline{AB}$ that is congruent to <3 and call it L_1, as shown.

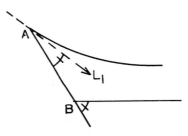

Now extend $\overline{AB}$ and $\overrightarrow{B\Omega}$ to form <4 as shown. Now by Theorem 4H, page 42, line L_1 and line $B\Omega$ do not intersect.

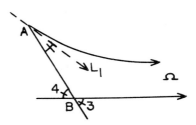

Now finish this part of the proof and then the whole proof.

Exercise 5-11   Draw a line L and a second line
                perpendicular to it, and mark
                off several points on the perpen-
                dicular, as shown.

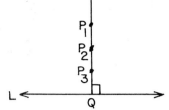

Sketch in the right-hand parallels to L through P_1, P_2,
and P_3. Now use Theorems 5E and D to explain why the
closer a point gets to L, the larger the angle of
parallelism becomes, though it never equals or becomes
larger than a right angle. To do this, you will have to
assume that if two lines are parallel to a third line in
the same direction, they are parallel to each other. Go
ahead and assume it. It can be proved.

Theorem 5F   If two infinite triangles are
              labeled as shown, and if $< B$
              $\cong < D$ and $\overline{AB} \cong \overline{CD}$, then
              $< A \cong < C$.

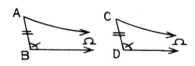

Proof: Prove this, using an indirect proof: Suppose $< A$ is greater
than $< C$, draw a line through A making an angle with $\overline{AB}$ that is congruent to $< C$. (Sketch all this as you read.) This line must intersect
$\overrightarrow{B\Omega}$. Call the point of intersection E. Mark off on $\overrightarrow{D\Omega}$ a segment
congruent to $\overline{BE}$. Call the endpoint F. Connect C to F, etc.)

63

<u>Definition of a Saccheri Quadrilateral</u>:
(See the definition of a quadrilateral, page 46.)

If two congruent segments are each drawn
perpendicular to some third line (on the
same side) as shown,

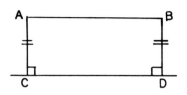

and their endpoints are joined by another
segment, the resulting figure is called
a Saccheri quadrilateral.
(In the figure, which is a Saccheri
quadrilateral, $\overline{AB}$ is called the <u>summit</u>,
$\overline{DC}$ is called the <u>base</u>, angles A and B
are called the <u>summit angles</u>.)

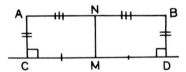

Learn this definition before continuing.

<u>Theorem 5G</u>   If a segment is drawn joining the <u>midpoints</u> of the base and
summit of a Saccheri quadrilateral, then that segment is
perpendicular to both the base and the summit.

<u>Proof</u>: The figure shown illustrates the
"If..." part of the theorem. Verify this
for yourself before continuing.

Now look back to page 26 at the definition of right angle.

Now draw $\overline{AM}$ and $\overline{BM}$ and prove the theorem. (This proof is the first use
of SSS, page 32.)

Use your sketch from Theorem 5G to prove the following theorem.

<u>Theorem 5H</u>   The summit angles of a Saccheri quadrilateral are congruent.

Exercise 5-12   Explain why the base and summit of a Saccheri quadrilateral are non-intersecting. (Hint: See exercise 5-10
and Theorem 4H.)

<u>Theorem 5J</u> The summit angles of a Saccheri quadrilateral are acute.

<u>Proof</u>: Starting with a Saccheri quadrilateral, as shown,

extend $\overline{CD}$ through D and draw the right-hand parallels to $\overline{CD}$ through A and B, as shown.

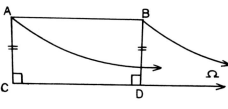

Now extend $\overline{AB}$ through B and number the angles, as shown.

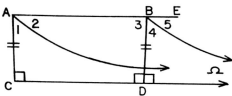

There are now at least three infinite triangles in the figure. Verify this for yourself before continuing.

The infinite triangles we are interested in are $\triangle AC\Omega$, $\triangle BD\Omega$, and $\triangle AB\Omega$.

What does Theorem 5F say about the relative sizes of angles 1 and 4? Write in your answer here before continuing.

What does Theorem 5E say about the relative sizes of angles 2 and 5? Answer here before continuing. (Hint: <5 is an exterior angle of infinite triangle $AB\Omega$.)

What does Theorem 5H say about angle 3 and the combination of 1 and 2, $<$ BAC?

How does the combination of angles 4 and 5 (< EBD) compare with the combination of angles 1 and 2 (< BAC)?

Now compare < EBD with < 3 and prove what was supposed to be proved.

What would an AAS Theorem about congruent triangles say? Make sketches to illustrate AAS triangles and then prove that they are congruent, using the method used in the proof of Theorem 4C, page 31.

<u>Theorem 5K</u> The angles of a triangle comprise angles which are less than two right angles.

Proof: Start with a triangle. We are to show that < 1 and < 2 and < 3 together are less than two right angles. (Recall that a triangle cannot have more than one obtuse angle.)

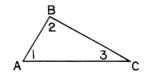

To do this, draw a line through the midpoints of $\overline{AB}$ and $\overline{BC}$, as shown. Call the midpoints M and N respectively. (If the triangle has an obtuse angle, the midpoints of its sides should be chosen for simplicity.)

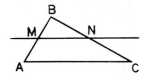

Now draw perpendiculars from A, B, and C to the line, as shown.

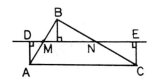

If it were true that $\overline{AD} \simeq \overline{EC}$, the quadrilateral ADEC would be a Saccheri quadrilateral (upside down). Is $\overline{AD} \simeq \overline{EC}$? Explain! (Use AAS twice.)

Now use Theorem 5J to finish the proof.

Exercise 5-13   Explain how Theorems 5D (page 60) and 5E (page 61) were used in the proof of Theorem 5K.

Exercise 5-14   Compare Theorem 5K with Theorem 4K on page 43. Are they in agreement? Are they contradictory?

Lobachevskian geometry is also called hyperbolic geometry, and the space it describes is called hyperbolic space. Here are some additional facts about hyperbolic space, which are interesting but not essential. If you don't wish to pursue the subject, you can skip to the reading list.

If our world were hyperbolic, a man could leave his house and go 5 blocks east and then 7 blocks north, and his wife, leaving the same house could go 7 blocks north first and then 5 blocks east, and they would not end up in the same place.

By the same token, you could draw horizontal and vertical axes. Then start at the origin and go 5 units to the right and 7 up. Then start again at the origin and go 7 up and then 5 over. You won't be at the same point as before!

The reason for this lies in the exotic properties of the quadrilateral with three right angles.

This figure (which does not exist in Euclidean space) is called a Lambert quadrilateral (4-lat) after one of the first men to be startled by its unusual properties.

The first unusual property of a 4-lat with three right angles is that the fourth angle must be acute.

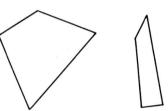

To see this, first consider a randomly drawn 4-lat, two of which are shown on the right.

Now a line drawn from any vertex to the opposite vertex of such a figure will divide the 4-lat into two triangles. The sums of the angles of both triangles are less than two right angles. Hence, the sum of the angles of any 4-lat are less than four right angles. From this fact, it is easy to see that if a 4-lat has three right angles, the fourth angle must be acute. (This fact can also be used to prove that, unlike Euclidean geometry, AAA is a congruence theorem for triangles.)

Even more surprising about a Lambert quadrilateral is the fact that the sides which comprise the acute angle are longer than the opposite sides. In the figure, $\overline{BD}$ is longer than $\overline{AC}$ and $\overline{AB}$ is longer than $\overline{CD}$.

To see this, we use an indirect argument. First, if $\overline{AC} \simeq \overline{BD}$, the figure would be Saccheri, and $\angle A$ would be acute, a contradiction. If $\overline{AC}$ were longer, mark off point E in such a way that $\overline{EC} \simeq \overline{BD}$. Now draw $\overline{EB}$. Do you see the contradiction? (See Theorem 4G, page 41.)

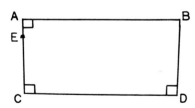

A similar argument proves that $\overline{AB}$ is longer than $\overline{CD}$.

Now perhaps you can see that the earlier statement about a man and his wife going east-north and north-east respectively, is so.

Now consider the familiar x-y axes. Starting at the origin and going over 3 and up 2 is not the same as going up 2 first and then over 3.

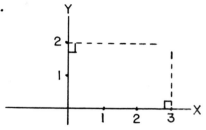

You can see that much of analytic geometry, trigonometry, and physics as we now know them would have to be rejected if we assumed that the world were Lobachevskian (hyperbolic) instead of Euclidean.

HISTORICAL BACKGROUND

The following is a list of readings which are intended to put Lobachevskian geometry (hyperbolic) into perspective for you and to pre-pare you for Riemannian (elliptic) geometry--System 6. (These two geom-etries are also called non-Euclidean.)

Begin by looking over the questions which follow the reading list so that you will have some questions in mind as you read. Then spend at least two hours on the reading.

The "hidden" intent is that you will see Geometry and Mathematics in general as a part of your cultural heritage, as a product of man's imagination.

(1)   The Spell of Mathematics, W. J. Reichmann (London: Methuen, 1967). Pages 195-216

(2)   A History of Astronomy, A. Pannekoek (New York: Interscience Pub., 1961). Page 487

(3)   The Search For Truth, E. T. Bell (New York: Reynal and Hitchcock, 1938). Pages 217-47

(4)   Men of Mathematics, E. T. Bell (New York: Simon and Schuster, 1937). Pages 294-306 and 484-505

(5)   Development of Mathematics, E. T. Bell (New York: McGraw-Hill Book Co., 1945). Pages 327-336

(6)   A History of Mathematics, Carl B. Boyer (New York: John Wiley & Son, 1968). Pages 585-590

(7)   The Mathematical Sciences, COSRIMS (Cambridge: M.I.T. Press, 1969). Pages 52-59

(8)   Mathematics in Western Culture, Morris Kline (New York: Oxford University Press, 1953). Pages 410-431

(9)   Mathematics and the Physical Universe, Morris Kline (New York: Crowell, 1959). Pages 443-463

(10)   The Main Stream of Mathematics, Edna E. Kramer (New York: Oxford University Press, 1955). Pages 241-262

(11)   Relativity For the Million, Martin Gardner (New York: Macmillan, 1962). Pages 85-105

(12) Relativity, the Special and General Theory, Albert Einstein
(Gloucester, Mass.: P. Smith, 1959). Pages 128 ff

(13) Mathematics--A Cultural Approach, Morris Kline (Reading, Mass.:
Addison-Wesley, 1962). Chapter 26

(14) Mathematical Thought From Ancient to Modern Times, Morris Kline
(New York: Oxford University Press, 1972). Chapter 36

(15) A Modern View of Geometry, Leonard M. Blumenthal (San Francisco:
W. H. Freeman, 1961). Chapter 1

(16) Mathematics in the Modern World, Morris Kline, editor (San
Francisco: W. H. Freeman, 1968). Chapter 16

(17) An Introduction to the Foundations and Fundamental Concepts of
Mathematics, Carroll Newsom & Howard Eves (New York: Holt,
Rinehart & Winston, 1965). Chapter Three

Questions:

1. Why is Euclid's 5th Postulate so much better known than any of the
   others?

2. What contributions did Karl Friedrich Gauss make to non-Euclidean
   geometry?

3. Was Lobachevsky really the first man to discover non-Euclidean
   geometry?

4. How is the Euclidean object called a "pseudo-sphere" related to
   Lobachevskian geometry?

TRANSITION

In changing from Euclidean geometry (System 4) to Lobachevskian
geometry (System 5), we merely dropped the phrase "but no more than
one" from the Axiom of Parallelism and a whole new world opened up!

The transition to Riemannian geometry (System 6) is also accom-
plished by a change in the axioms, but there are a few more changes
than simply an alteration of the Axiom of Parallelism. The changes are
simple but the effects are profound.

System 6

riemannian geometry

System 6 is called Riemannian geometry after Bernard Riemann, it is also called elliptic geometry.

The words "point," "line," "plane," "congruent," and "between" are primitive. The definitions of System 4 all carry over to System 6, except the definition of "parallel lines."

AXIOMS OF CONNECTION

1. For any two distinct points, there is at least one line through them both.

2. Two lines cannot intersect in more than two points.

3. There are at least two points on every line.

4. There are at least three points on every plane, not all of which are on the same line.

5. For any (every) three distinct points that do not lie on the same line, there is exactly one plane on all three.

6. If two points of a line lie on a plane, then the entire line lies on that same plane.

7. There are at least four points, not all of which lie on the same plane.

AXIOMS OF ORDER

1. For any two points of a line, there exists at least one third point of the line which lies between them.

2. Every segment has exactly one midpoint.

MEASURE AXIOM

For every segment, there is a positive (real) number called the <u>length</u> <u>of the segment</u>. For any (every) two segments, the shorter segment has the smaller length. For any positive (real) number, there is a segment which has that number as its length. Congruent segments have the same length.

AXIOM OF SEPARATION

Every line partitions every plane in which it lies into two half-planes (called the left and right half-planes, respectively) in such a way that

(1) Exactly one of the following is true; every point of the plane is in the left half-plane or in the right half-plane or on the line.

(2) For any point in one half-plane and any point in the other, the line joining them intersects the line which formed the half-planes

(3) If two points (call them A and B) lie in the same half-plane, then there is a segment having A and B as its endpoints, every point of which lies in the half-plane with A and B.

AXIOMS OF CONGRUENCE

1. If AB is a segment of a line L and C is a third point on the same or some other line, then there is on either side of C and on the same line exactly one fourth point D which will make AB congruent to CD.

2. Every segment is congruent to itself. If any segment is congruent to a second segment, then the second segment is congruent to the first.

3. If seg AB is congruent to seg CD and seg CD is congruent to seg EF, then seg AB is congruent to seg EF.

4. If L is any line and P is any point of L, then there is exactly one line through P which is perpendicular to L.

5. Suppose ABC is any angle. If O is a point on some line L and M is a second point on L then (on either side of L) there is exactly one ray with its endpoint at O which forms an angle with ray OM which is congruent to angle ABC.

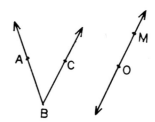

6. Every angle is congruent to itself. If any angle is congruent to a second angle, then the second angle is congruent to the first.

7. If angle ABC is congruent to angle POR and angle POR is congruent to angle XYZ, then angle ABC is congruent to angle XYZ.

8. If there are two triangles ABC and XYZ and if AB is congruent to XY and angle ABC is congruent to angle XYZ and BC is congruent to YZ, then triangle ABC is congruent to triangle XYZ.

(Also stated: If two sides and the included angle of one triangle are congruent respectively to two sides and the included angle of another, the triangles are congruent.) This axiom is abbreviated SAS.

AXIOMS OF TRICHOTOMY

1. For any two segments, the first is either shorter than, congruent to, or longer than the second.

2. For any two angles, the first is either smaller than, congruent to, or larger than the second.

FUNDAMENTAL ELLIPTIC AXIOM

If L is any line and if P is any point which is not on L, then there are no lines in the same plane with L and P which pass through P and do not intersect L. (This is another way of saying "all lines intersect.")

There are three more elliptic axioms which will be given soon. They aren't included above because they use a concept which hasn't been developed yet.

In the set of elliptic axioms given so far, there are eight changes from the axioms for Euclidean geometry. See how many of them you can find before continuing.

Exercise 6-1 Which theorems from Euclidean geometry and Lobachevskian geometry are still valid in Riemannian geometry so far?

PRELIMINARIES

Theorem 6A If L is any line and if two other lines are perpendicular to L and in the same plane, then the two other lines will intersect.

Proof: Here is a sketch of the "If..." part of the theorem.

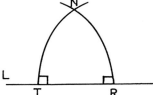

Now use the Fundamental Elliptic Axiom to explain why the theorem is true.

Draw a picture of the result in the space below.   (Look again at meta-axiom 1, page 8.)

The picture you were just asked to draw should look something like this

N

L

T   R

Is every exterior angle greater than the non-adjacent interior angles?

<u>Theorem 6B</u>   Given the above situation, prove that $\overline{TN} \simeq \overline{RN}$.

Proof: The proof is indirect.   Suppose
$\overline{TN} > \overline{RN}$.   Then, as shown on the right, mark off on $\overline{TN}$ a point W in such a way that $\overline{TW} \simeq \overline{RN}$.

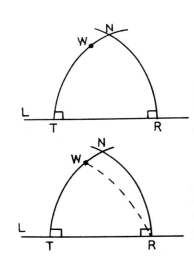

Now connect W to R, as shown.   (The fact that there may be more than one such segment won't change this proof.)

74

Now notice $\triangle$TWR and $\triangle$TNR. $\overline{TW} \simeq \overline{RN}$, $< WTR \simeq < NRT$, and $\overline{TR}$ is congruent to itself. Find the contradiction and finish the proof.

Notice that the proof does not depend upon angles NTR and NRT being right angles, but only upon the fact that they are congruent.

Call the point of intersection of the two perpendiculars N.

<u>Theorem 6C</u> If a third line is drawn perpendicular to L in the plane with L and N, then it will also pass through N.

<u>Proof:</u> We have to show that a line drawn perpendicular to L at <u>any</u> point will pass through N. We do this by a process of slow buildup.

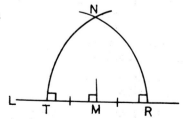

Choose a point Q on the other side of R in such a way that $\overline{TR} \simeq \overline{RQ}$. Then draw a perpendicular to L at Q, as shown. This perpendicular must pass through N. Use an indirect proof and ASA to show this.

Now draw a perpendicular to L through the midpoint M of $\overline{TR}$, as shown.

Suppose this perpendicular does not pass through N. It must intersect both $\overline{TN}$ and $\overline{RN}$ (or their extension). One such possibility is shown. Use ASA to conclude the contradiction that $\overline{WM} \simeq \overline{UM}$. Other possibilities will lead to similar contradictions.

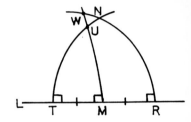

Now if $\overline{TR}$ is divided into three equal segments by points Q and P and perpendiculars to L at Q and P are drawn as shown, a similar though somewhat more complex argument would show that these perpendiculars also pass through N.

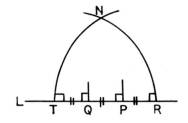

If $\overline{TR}$ is divided into any number of equal segments (as shown), a similar but increasingly complex argument could be used to show that all the perpendiculars to L through these points pass through N. The same thing could be done between R and Q (where $\overline{TR} \simeq \overline{RQ}$), etc.

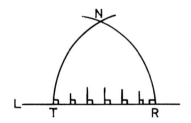

For points of L not accounted for, a process from the calculus must be used. Please consider that done, for the sake of progress. If you wi the proof is finished.

Note that <u>all</u> the segments from N perpendicular to L are congruen to each other, even though they may not look like it. Since "congruen means that (intuitively) the segments will fit over each other exactly if they are moved together, it appears that moving a segment (in the elliptic world) changes its shape. How do you account for this?

Remember, all the lines of this system are "straight," though thi may require you to change your conception of straightness.

Look again at meta-axiom 1, page 8.

Theorem 6D If a third line passes through N in the plane with L and N then it must be perpendicular to L. (Before continuing, compare this theorem with Theorem 6C.)

<u>Proof</u>: Look at the picture on the right. It agrees with the "If" part of the theorem.

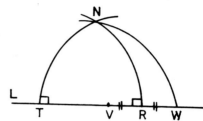

To see why the theorem is true, use a direct proof. Extend the third line till it intersects L. Call the point of intersection W, and mark off on the other side of R a segment congruent to $\overline{RW}$. Call its endpoint V, as shown. (V may or may not fall between T and R, but this won't affect the proof.)

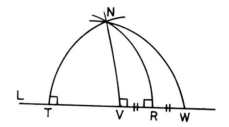

Now draw a line through V perpendicular to L. By Theorem 6C, this line must pass through N.

Finish the proof.

Read the following carefully!

Summarizing Theorems 6B, 6C, and 6D, we can say that for any (every) line L there is a point N which has the property that any line perpendicular to L passes through N and any line passing through N is perpendicular to L. All the segments from N perpendicular to L are the same length.

<u>Definition</u>: The point N is called a <u>pole</u> of the line L.

<u>Theorem 6E</u> The length of a segment from a line to its pole is the same as the length of a segment from any other line to <u>its</u> pole.

<u>Proof</u>: Draw two lines and their poles, as shown on the right.

Now mark off <u>congruent</u> segments on the two lines, as shown, i.e., make $\overline{AB} \simeq \overline{CD}$.

77

Now draw perpendiculars to L_1 and L_2 at A, B, C, and D.   Then use Theorem 6C and ASA to finish the proof.

Definition: The <u>distance from a point to a line</u> is the length of a perpendicular segment from the point to the line.

Definition: The distance from a line to its pole is called the <u>polar distance</u> and is denoted by the letter "r."

From Theorem 6E we know that r is a constant, not a variable. (If elliptic geometry is the geometry of the universe, then r is astronomically large!*)

THE "OTHER SIDE" OF THE LINE

Beginning with the picture on the right, let's investigate what happens below L.

If $\overline{TN}$ and $\overline{RN}$ are extended through T and R respectively, we cannot know whether they will intersect again.   (They will, but it requires proof.)

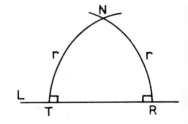

Instead, extend $\overline{TN}$ through T to a length r and call the endpoint S, as shown.

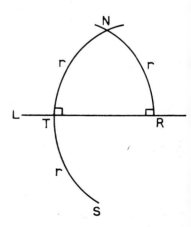

---
* See Chapter XXXI of <u>Relativity: The Special and General Theory</u> (15th ed.) by Albert Einstein.   (New York: Crown Pub., Inc., 1952)

Now connect S and R. (Do not assume that $\overline{SR}$ and $\overline{NR}$ are part of the same line or that $\overline{SR} \perp$ L.)

$$\triangle\,NTR \cong \triangle\,STR \text{ by SAS}$$

Hence $<$ TRS is a right angle and also $\overline{NR}$ and $\overline{RS}$ are segments of the same line because of Axiom 4 of Congruence.

Arguments similar to those of Theorems 6B, 6C, and 6D show that S is also a pole of L. Convince yourself of this before continuing.

The distance between two poles of the same line is 2r.

AXIOM OF INTERSECTION

For any (every) two distinct points which are not <u>both</u> poles of the same line, there is no more than one straight line through them both.

AXIOM OF BETWEENNESS

For any (every) three points of a segment of length less than 4r, exactly one lies between the other two.

AXIOM OF PERPENDICULARITY

If L is any line and P is any point which is not a pole of L, then there is exactly one line through P perpendicular to L.

Exercise 6-2 Why couldn't the above axioms have been given sooner?

<u>Theorem 6F</u> Any two lines will intersect in two points. (Look at Axiom 2 of Connection.)

<u>Proof:</u> The Fundamental Elliptic Axiom says that any two lines must intersect at least once. So let L_1 and L_2 be any two lines and call their point of intersection A, as shown on the right.

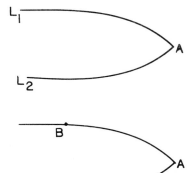

Now starting at A, mark off on both L_1 and L_2 a distance equal to "r" and call the points so determined B and C, as shown on the right.

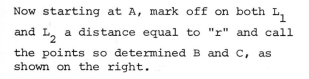

Draw a line through B perpendicular to L_1, explain why it must pass through C, and then finish the proof. (Suppose the perpendiculars at B and C are not the same line. Draw a line from A to their point of intersection.)

Exercise 6-3   Explain why every line has a length equal to 4r. (Hint: Let L be a line and A a point on L. Mark off segments of length r on either side of A along L. Call the endpoints P and P'. Draw a second line through A perpendicular to L. Now use Theorem 6F and Axioms 4 of Congruence and 2 of Connection.)

Exercise 6-4   Use the results of exercise 6-3 to explain why the defini- tion of "$\overline{AB}$ is shorter than $\overline{CD}$" (page 29) is no longer appropriate. Correct the flaw by putting appropriate limits on the lengths.

Note that from now on given any three points on a line, it will be necessary to specify which one is to be considered "between" the other two. Also, any two points which are not poles of the same line deter- mine two segments (instead of one, as in the other systems), one of length less than 2r and one of length greater than 2r.

To say that a line has a length is startling! In school most of us learned that lines went off in both directions to infinity. Segments of lines had length, but lines were infinite. This is still true in Euclidean and hyperbolic geometry.

We can now explain why the proof of Theorem 4G (the exterior angle of a triangle is greater than either of the two interior angles which are not adjacent to it) is no longer valid.

To prove that <4 > <1 we picked the midpoint of $\overline{AB}$, called it M, con- nected M to C and then extended $\overline{CM}$ through M in such a way that $\overline{DM} \simeq \overline{MC}$. This was OK in Euclidean and hyper- bolic geometry, because we assumed (without actually saying so) that lines are infinite in extent.

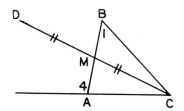

In elliptic (Riemannian) geometry, however, it is possible that the length of $\overline{DC}$ is greater than 4r, in which case D might actually end up somewhere between C and M. Obviously, if this happened the rest of the proof wouldn't work.

The proof is still valid (at least) if the lengths of $\overline{BC}$ and $\overline{AC}$ are less than r, (since this would make $\overline{DC}$ less than 2r) and also at other times. It is an interesting exercise to determine the conditions on the lengths of the sides under which the theorem is still true.

ANOTHER LOOK AT TRIANGLES

Theorem 6G   If a triangle has one right angle, then the other two angles are either acute, right, or obtuse according as the side opposite them is less than r, equal to r, or greater than r, respectively.

Proof: Referring to the figure on the right, explain why the following six statements are true:

(1)   If < B is a right angle, $\overline{AC}$ has length r.

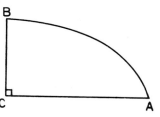

(2)   If $\overline{AC}$ has length r, then < B is a right angle.

(3)   If < B is acute, then the length of $\overline{AC}$ is less than r. (To see this, draw the perpendicular to $\overline{CB}$ at B and investigate where it intersects $\overline{AC}$ or the extension of $\overline{AC}$.)

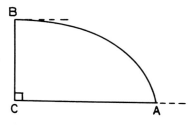

(4) If the length of $\overline{AC}$ is less than
     r, then < B is acute. (To see
     this, extend $\overline{AC}$ through A a
     distance equal to r and connect
     the endpoint with B.)

(5) If < B is obtuse, then the length
     of $\overline{AC}$ is greater than r. (Use
     an indirect proof.)

(6) If the length of $\overline{AC}$ is greater
     than r, then < B is obtuse.

Similar arguments could be made using < A with side $\overline{BC}$.

Exercise 6-5   If all the angles of a triangle are acute or if two are
               acute and one is a right angle, explain why the sides of
               the triangle are all of length less then 2r.

        The previous theorem will be used in the next group of theorems.
Before continuing, review the definition of a Saccheri quadrilateral
on page 64.

Theorem 6H   The line joining the midpoints of the base and summit of a
             Saccheri quadrilateral is perpendicular to both the base
             and the summit.

                The proof of this theorem in System 5 (Theorem 5G, pa
             64) used SAS, SSS, and the definition of perpendicular
             lines. Since these three things also hold true in System
             the theorem carries over from hyperbolic to elliptic geom-
             etry. It's not necessary to prove it again.

        In Euclidean and hyperbolic space, a quadrilateral can be drawn a
large as we please, but in elliptic space, the nature of the space itself requires that some figures can only be drawn so large and no larg
The following theorem illustrates this.

82

<u>Theorem 6J</u>   The two congruent sides of a Saccheri quadrilateral have a
length less than r, the polar distance, and the base and
summit have a length less than 2r.

<u>Theorem 6K</u>   The summit angles of a Saccheri quadrilateral are congruent
and obtuse.

<u>Proof:</u>   The proof that the summit angles are
congruent carries over from Theorem 5H, page
64.

To prove that angles 1 and 2 are obtuse,
join the midpoints of the summit and base,
as shown on the right.

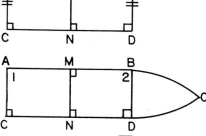

Now extend $\overline{AB}$ and $\overline{CD}$ through B and D
respectively until they meet at a point
O, as shown on the right.

Then point O must be the pole of line MN, so the length of $\overline{NO}$ is equal
to r.

Now look at triangle BDO.   The length of $\overline{DO}$ is less than r.   < BDO is
a right angle.   Use Theorem 6G to finish the proof.

<u>Theorem 6L</u>   The angles of a triangle comprise angles which are greater
than two right angles.

<u>Proof:</u>   If the triangle has at least two right angles or at least two
obtuse angles or a right and an obtuse angle, the theorem is obvious.
The only other possibility is that a triangle might have at most one
angle which is either right or obtuse.   Thus it has at least two acute
angles.

<u>Case 1:</u>   Suppose the third angle is a right angle.
(Look again at exercise 6-5.)   Draw the triangle
with right angle at C, as shown, pick the mid-
points of $\overline{AC}$ and $\overline{BC}$ and proceed as in the
analogous proof in System 5, Theorem 5K, page 66.

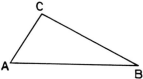

<u>Case 2:</u>   The third angle is not a right angle.   Complete the proof.
(Hint: Draw a line from the vertex of the third angle perpendicular
to the opposite side.)

This is as far as this book goes in System 6.

In order to catch the idea out of which this book has grown, the reader now should compare and contrast the last theorem proved in each of Euclidean, hyperbolic, and elliptic geometries: Theorem 4K, page 43; Theorem 5K, page 66; and Theorem 6L above.

CONCLUSION

There are a couple of important questions which people frequently ask. Here they are along with some ideas which might point the way toward some discussion.

1. We only changed some of the axioms of System 4. Why not change some others? Why not change them all? How do you know which ones to change?

There are three standards or criteria which one is supposed to use when choosing a set of axioms to work with.

(a) An axiom set must be consistent. This means that one must not be able to start with some set of axioms and prove two theorems or statements which contradict each other. How, then, can we be sure that a thousand mathematicians working for a thousand years on a certain system (for example, System 4) will never come up with two contradictory theorems? The answer is, we can't be sure! This is meta-axiom 6, page 47.

It has been proved that if someone should stumble across a logical flaw in one of the three geometries, similar flaws will exist in the other two even though we can't find them. In other other words, though it is impossible to tell if any of the three geometries (Euclidean, elliptic, hyperbolic) is consistent (within itself), they can be shown to be relatively consistent (with respect to each other).*

---

*See Chapter VIII of Introduction to Non-Euclidean Geometry by Harold E. Wolf (New York: Holt, Rinehart & Winston, 1945).

(b) Each of the axioms in the set must be <u>independent</u>. This means that, theoretically at least, no axiom should be provable using the others. In effect, this also means that if you want to use a fact and you know for sure that it can't be proved, then you must assume it, i.e., make it an axiom. How can you be sure that some fact can't be proved? You can't always be sure. This is meta-axiom 7, page 47.

(c) An axiom set should be <u>complete</u>. This means that any statement about the objects and relations of the system should be capable of being either proved or disproved on the basis of the axioms. This is meta-axiom 4, page 18.

It is impossible to prove that any set of axioms meets the requirements of meta-axiom 4. That this is so was demonstrated by the logician, Kurt Gödel. The famous "Gödel's Theorem" asserts that it is impossible to tell whether or not an axiom system like our Systems 4, 5, or 6 is complete. If you wish to pursue this, see the article on Godel in James R. Newman's <u>World of Mathematics</u> in any library.

These three criteria are esthetic, not logical.

2. Which geometry is really true?

There is no answer for this question, because we don't know what "really" means.

In a logical sense, they are all equally true (or equally false), since logical truth depends on the criteria given above for axiom sets and on the rules of logic you use.

In an everyday "real world" sense, it is impossible to say which one is true. The space we move in appears to be Euclidean, but then the instruments we use to measure it (yardsticks, meters, etc.) are not exact and are based on Euclidean principles anyway.

It may be that the universe of the Milky Way is indeed elliptic or hyperbolic, but the measurements we make on earth are so small in relative magnitude that what we take for Euclidean is really elliptic or hyperbolic, but the difference is too small to detect.

For a further discussion of geometry and reality, see the opening chapters of Einstein's little book on relativity mentioned in the last reading list, or the last few pages in Chapter D of <u>Number: The Language of Science</u> by Tobias Dantzig (New York: Macmillan, 1954).

Appendices

A <u>ratio</u> is simply a quantity expressed as one number over (divided by) another. Thus any common fraction (proper or improper) is a ratio, for example, $\frac{2}{3}$ or $\frac{5}{4}$. Whole numbers such as 5 or 92 can be thought of as ratios, since they can be written as $\frac{5}{1}$ and $\frac{92}{1}$ (if it's conveni-ent). Other examples of ratios are $\frac{\pi}{2}$ or $\frac{\sqrt{7}}{19}$.*

The ratio $\frac{2}{3}$ may be read "the ratio of 2 to 3," $\frac{5}{4}$ may be read "the ratio of 5 to 4." etc.

A <u>proportion</u> is simply two ratios set equal to each other. The following are examples of proportions:

(1) $\quad \frac{4}{3} = \frac{8}{6}$ $\qquad\qquad$ (2) $\quad \frac{2x}{5} = \frac{9}{10}$ $\qquad\qquad$ (3) $\quad \frac{7}{4} = \frac{5}{2}$

If the two ratios are, in fact, equal as in the first example, the pro-portion is called <u>valid</u>. If the two ratios are not equal, as in the third example, the proportion is called <u>invalid</u>.

Axiom of Proportions

A proportion of the form $\frac{a}{b} = \frac{c}{d}$ is valid if and only if the product "a times d" equals the product "b times c," that is, if and only if $ad = bc$.

The proportion is read "a is to b as c is to d."

---

* Every rational number, being one integer over another, is a ratio, but not all ratios are rational numbers. π and $\sqrt{7}$ are not integers.

The above axiom introduces the concept often called "cross-multiplying."

$$\frac{a}{b} = \frac{c}{d}$$

$$ad = bc$$

Thus, to check the validity of a proportion, the Axiom of Proportions requires one to cross-multiply.

Example 1 Read and check the validity of the proportion $\frac{4}{16} = \frac{2}{8}$.

Solution: 4 is to 16 as 2 is to 8

4 x 8 = 32 and 16 x 2 = 32, so the proportion is valid because

4 x 8 = 16 x 2.   (Also, both fractions in the proportion

reduce to $\frac{1}{4}$.)

Checks for validity should be done mentally when possible.

Example 2 Read and check the validity of $\frac{7}{17} = \frac{21}{50}$.

Solution: 7 is to 17 as 21 is to 50.

7 x 50 = 350 and 17 x 21 = 357, so the proportion is not valid because 7 x 50 ≠ 17 x 21.

Example 3 Find a replacement for x so that $\frac{7}{17} = \frac{21}{x}$ will be a valid proportion.

Solution: Cross-multiplying, we get

$$7x = 17(21)$$

$$x = \frac{17(\overset{3}{\cancel{21}})}{\underset{1}{\cancel{7}}}$$

$$x = 51$$

Exercise A-1 Read the following proportions and check them for validity. If the proportion contains a variable, find the replacement for the variable which will make the proportion valid.

Answers are provided (page 90). If you know any short-cuts, use them.

$$(1) \quad \frac{13}{9} = \frac{117}{80} \qquad\qquad (2) \quad \frac{19}{14} = \frac{38}{28}$$

$$(3) \quad \frac{x}{3} = \frac{9}{6} \qquad\qquad (4) \quad \frac{2}{3y} = \frac{8}{12}$$

$$(5) \quad \frac{4}{5} = \frac{12}{15} \qquad\qquad (6) \quad \frac{4}{15} = \frac{5}{12}$$

<u>Definition of Similar Triangles</u>: <u>Similar triangles</u> are triangles whose corresponding angles are congruent and the lengths of whose corresponding sides form three valid proportions.

Here is an example of similar triangles:

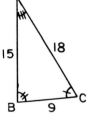 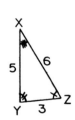

$\overline{AB}$ and $\overline{XY}$ are corresponding sides and are in the ratio of 15 to 5, $\frac{15}{5}$.

$\overline{BC}$ and $\overline{YZ}$ are corresponding sides and are in the ratio of 9 to 3, $\frac{9}{3}$.

$\overline{AC}$ and $\overline{XZ}$ are corresponding sides and are in the ratio of 18 to 6, $\frac{18}{6}$.

Any two of the three ratios will form a valid proportion, i.e.,

$$\frac{15}{5} = \frac{9}{3} \quad \text{or} \quad \frac{15}{5} = \frac{18}{6} \quad \text{or} \quad \frac{9}{3} = \frac{18}{6}$$

so the sides are "in proportion" and the triangles are similar.

Intuitively, similar triangles have the same shape but not the same size. Congruent triangles are just a special kind of similar triangles.

The concept of similarity has meaning only in Euclidean geometry. The only similar triangles in hyperbolic geometry are congruent.

This definition of similar triangles is the mathematical equivalent of using a shotgun to kill a fly. It is inelegant because it says too much. Given the proper axioms and definitions, it is possible to <u>prove</u> that if the corresponding sides of two triangles form three valid proportions, then the corresponding angles must be congruent, and vice versa. We will assume that this statement is true.

<u>Examples</u> Determine whether or not the triangles are similar. If so, find the lengths of unmarked sides and the size of all angles, if possible.

(1)

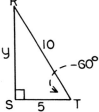

Since the length of each side of the larger triangle is 3 times the length of the corresponding side of the smaller, the triangles are similar.

$$(\frac{24}{8} = \frac{21}{7} \quad \text{and} \quad \frac{24}{8} = \frac{18}{6}$$

and $\frac{21}{7} = \frac{18}{6}$. Each of these ratios

reduce to 3, so the proportions are valid.)

With what we've had so far, it is not possible to find the size of the angles.

(2)

The triangles are similar, since corresponding angles are congruent.

(a) To find the length of $\overline{AC}$, 10 is to x as 5 is to 3:

$$\frac{10}{x} = \frac{5}{3}$$

$$5 x = 30$$

$$x = 6$$

(b) To find the length of $\overline{RS}$, y is to $3\sqrt{3}$ as 5 is to 3:

$$\frac{y}{3\sqrt{3}} = \frac{5}{3}$$

$$3y = 15\sqrt{3}$$

$$y = 5\sqrt{3}$$

(You could also say y is to $3\sqrt{3}$ as 10 is to 6 or $3\sqrt{3}$ is to y as 3 is to 5, etc.)

(c)   < R ≃ < A and < T ≃ < C: < R = 30° and < C = 60°.

(3)

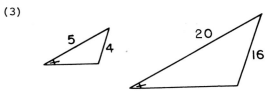

There is not enough information to tell whether or not the triangles are similar.

Exercise A-2   Determine whether or not the triangles are similar. If so, find the length of the unmarked sides and the size of all angles, if possible.

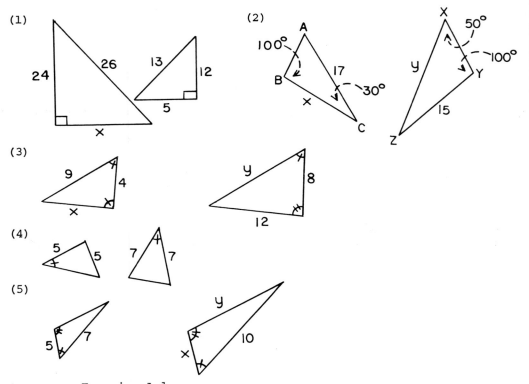

(1)

(2)

(3)

(4)

(5)

Answers: Exercise A-1

(1)   13 is to 9 as 117 is to 80   (2)   19 is to 14 as 38 is to 28
       not valid   valid

(3)   x is to 3 as 9 is to 6   (4)   2 is to 3y as 8 is to 12
       x = 4-1/2   y = 1

(5)   4 is to 5 as 12 is to 15   (6)   4 is to 15 as 5 is to 12
       valid   not valid

Exercise A-2

(1) x = 10

(2) The triangles are similar, since corresponding angles are congruent. However, there is not enough information to find the lengths of the sides.

(3) x = 6, y = 18

(4) Even though both are isosceles, the triangles are not necessarily similar. If all the angles are the same size, however, (a possibility) then the triangles are similar, the angles are all 60°, and the two unmarked sides are 5 and 8.

(5) x = 7-1/7, y cannot be determined.

APPENDIX B: SQUARE ROOTS AND THE PYTHAGOREAN THEOREM

Often students have no intuitive feel for the magnitude of irrational numbers like $\sqrt{3}$ or $\sqrt{20}$. Since these numbers will be used for the Pythagorean Theorem, the following is intended as an aid in overcoming this problem.

Start at the top of the first column below and read to the bottom. Then go to the top of the second column and read down, then to the top of the third, etc. The purpose is to familiarize you with the table.

$\sqrt{1}$ = 1	$\sqrt{12}$	$\sqrt{23}$	$\sqrt{34}$
$\sqrt{2}$	$\sqrt{13}$	$\sqrt{24}$	$\sqrt{35}$
$\sqrt{3}$	$\sqrt{14}$	$\sqrt{25}$ = 5	$\sqrt{36}$ = 6
$\sqrt{4}$ = 2	$\sqrt{15}$	$\sqrt{26}$	$\sqrt{37}$
$\sqrt{5}$	$\sqrt{16}$ = 4	$\sqrt{27}$	$\sqrt{38}$
$\sqrt{6}$	$\sqrt{17}$	$\sqrt{28}$	--
$\sqrt{7}$	$\sqrt{18}$	$\sqrt{29}$	--
$\sqrt{8}$	$\sqrt{19}$	$\sqrt{30}$	--
$\sqrt{9}$ = 3	$\sqrt{20}$	$\sqrt{31}$	
$\sqrt{10}$	$\sqrt{21}$	$\sqrt{32}$	
$\sqrt{11}$	$\sqrt{22}$	$\sqrt{33}$	

Notice that $\sqrt{3}$ is between $\sqrt{1}$ which is 1 and $\sqrt{4}$ which is 2 and closer to $\sqrt{4}$ than to $\sqrt{1}$. As an estimate (which is all we need) $\sqrt{3}$ is about 1.7. Also, $\sqrt{20}$ is between 4 and 5 (that is, between $\sqrt{16}$ and $\sqrt{25}$) and is closer to 4 (or $\sqrt{16}$) than to 5, so we could estimate that $\sqrt{20}$ is about 4.4.

Use the table to estimate the following numbers the way $\sqrt{3}$ and $\sqrt{20}$ were estimated above. Use the answers to help you by checking each answer as you go.

(1) $\sqrt{7}$

(2) $\sqrt{14}$

(3) $\sqrt{31}$

Check your answers.

(4) $\sqrt{18}$

(5) $\sqrt{27}$

(6) $\sqrt{70}$

(7) $\sqrt{94}$

(8) $\sqrt{5}$

Answers   (Your decimal might be off 1 either way.)

(1) 2.7   (2) 3.8   (3) 5.6   (4) 4.2   (5) 5.1   (6) 8.3   (7) 9.8   (8) 2.2

The most famous mathematical theorem, the Pythagorean Theorem, is named after Pythagoras, the Greek mathematician and mystic of the 6th century B.C.

The theorem, a theorem of Euclidean geometry, deals exclusively with right triangles. It says that the <u>square</u> of the length of the hypotenuse of any right triangle is equal to the <u>sum</u> of the <u>squares</u> of the lengths of the other two sides. (The hypotenuse is the side opposite the right angle.)

For example, in the figure on the right, the length of $\overline{AB}$, squared, equals the square of the length of $\overline{AC}$ plus the square of the length of $\overline{BC}$. In symbols this would be written $(AB)^2 = (AC)^2 + (BC)^2$

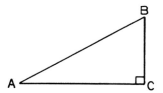

Example 1

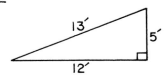

$13^2 = 169$

$12^2 = 144$

$5^2 = 25$

so $13^2 = 12^2 + 5^2$

or $169 = 144 + 25$

92

Example 2

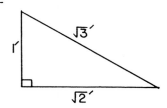

$(\sqrt{3})^2 = 3$

$(\sqrt{2})^2 = 2$

$1^2 = 1$

so $(\sqrt{3})^2 = (\sqrt{2})^2 + 1^2$

or $3 = 2 + 1$

Example 3

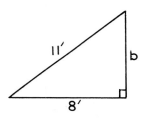

Find the length a.

Solution: $a^2 = 4^2 + 5^2$

$= 16 + 25$

$= 39$

so $a = \sqrt{39}$ or about 6.2"

Example 4

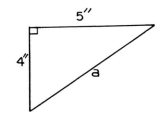

Find the length b.

Solution: $11^2 = b^2 + 8^2$

$121 = b^2 + 64$

$b^2 = 121 - 64$

$= 57$

so $b = \sqrt{57}$ or about 7.6'.

Exercise B-1   Find the length of the missing side. Use the answers to
help you by checking each answer as you go. (Do as many
mentally as possible.)

(1)

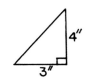

(2)

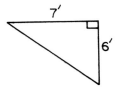

(3)

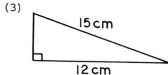

(4)

(5)   The triangles are similar.   Find x.

93

(6) Find the length of $\overline{BD}$. (Hint: First find the length of $\overline{DC}$.)

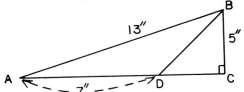

(7) The length of $\overline{BE}$ is 1/3 the length of $\overline{BA}$.
The length of $\overline{DB}$ = 10 cm. The length of $\overline{CA}$ is 20 cm. Find the
length of $\overline{BA}$.

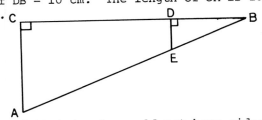

(8) Explain why a right triangle could not have sides of lengths shown
in the figure.

Answers

(1) 5"

(3) 9 cm

(5) x = 10

(7) $\sqrt{1300}$ = $\sqrt{13}$ x $\sqrt{100}$ or
     about 37 cm

(2) $\sqrt{85}$ or about 9.2'

(4) $\sqrt{96}$ or about 9.8

(6) $\sqrt{50}$ or about 7.1"

(8) $12^2 \neq 10^2 + 8^2$ (144 $\neq$ 164)

       To prove the Pythagorean Theorem, we will give a geometrical and
intuitive proof of great elegance. (There are over two dozen different
proofs now known.)

       Start with the triangle ABC as shown
on the right, with c the length of the side
opposite <C, b the length of the side
opposite <B, and a the length of the side
opposite <A.

We are supposed to prove that $a^2 + b^2 = c^2$.

Geometrically, the number a^2 can be interpreted as the <u>area</u> of a square whose sides are "a" units on a side, like Square 1 in the figure.

Similarly, b^2 is the area of Square 2, and c^2 is the area of Square 3.

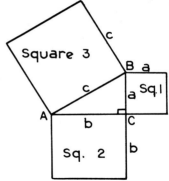

Proving that $a^2 + b^2 = c^2$ is equivalent to proving that the area of Square 1 plus the area of Square 2 equals the area of Square 3.

To do this, draw Square 3 on the other side of $\overline{AB}$, as shown.

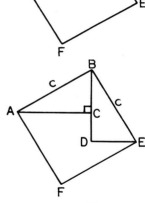

Now extend $\overline{BC}$ through C to a point D in such a way that $\overline{BD} \simeq \overline{AC}$, and connect D to E, as shown.

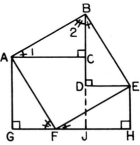

Now $<EBD \simeq <BAC$ because both are complementary to $<ABC$. Thus $\triangle ABC \simeq \triangle BDE$ (SAS), and so $<BDE$ is a right angle and $\overline{DE} \simeq \overline{BC}$.

From F, draw two segments, one making an angle with $\overline{FE}$ congruent to <1 and one making an angle with $\overline{FA}$ congruent to <2.

Now complete triangles FEH and FGA as shown and extend $\overline{BD}$ till it intersects $\overline{FH}$ at J, as shown.

A few moments reflection will show that $\triangle FEH \simeq \triangle FGA$ and GFH is one line.

Now complete the proof by showing that Square ACJG plus Square DEHJ equals Square ABEF.

APPENDIX C: FORMULAS

Listed below are some of the mensuration formulas from Euclidean geometry, which are useful in mathematics generally and in many sciences. (The number π may be approximated by 3.14 or 3.1416.)

1. Circle

Area equals pi times the square of the radius:

$$A = \pi r^2$$

Circumference equals pi times the diameter:

$$C = \pi d$$

(Since the diameter is twice the radius, this formula can be written $C = 2\pi r$.)

2. Square

Area equals the square of the length of one side:

$$A = s^2$$

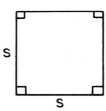

3. Rectangle

Area equals length times width:

$$A = lw$$

Perimeter equals twice the length plus twice the width:

$$P = 2\left(l\right) + 2w$$

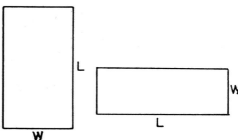

4. Triangle

Area equals one-half the base times the height:

$$A = 1/2\ b\ h$$

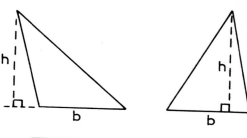

5. Parallelogram

Area equals base times height:

$$A = b\ h$$

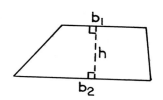

6. Trapezoid

Area equals one-half the height times the sum of the bases:

$$A = 1/2\ h(b_1 + b_2)$$

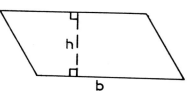

SOLUTIONS TO EXERCISES

System 1

Exercise 1-3   1.   Each pair of straight lines has exactly one point in common.

2.   Each point is on exactly two straight lines.

3.   There are exactly four straight lines.

Exercise 1-4   Rule 3

Exercise 1-5   Using model 1, suppose for instance that member A is removed. Then we would have:

But now the two committees on top have no member in common, which is a violation of Rule 1. Therefore member A can't be removed. No other member can be removed for a similar reason.

Therefore there cannot be five members.

Exercise 1-6 Using model 1, if there were a seventh member (call him G), Rule 2 says that he would have to be on two committees. Suppose we put him in the bottom two committees, like this:

| A, B, C | | A, D, E |

| B, D, F, G | | C, E, F, G |

Now those two committees have both F and G in common, and that violates Rule 1. Similarly, no matter what two committees you put G on, those two committees are going to have two members in common, violating Rule 1.

Therefore, there cannot be seven members.

Exercise 1-7 There are six members.
There are three members on each committee.
(Other statements are possible.)

System 2

Exercise 2-1 axiom 3

Exercise 2-2 By axiom 3 there is one committee. By axiom 4, there are three members on it.

Exercise 2-3 By axiom 5, there is at least one more member who is not on the first committee, so there are at least four members. Must there be another committee for the fourth member? It is not obvious that the fourth member has to be on a committee at all! Axiom 1 can be used to show that he does, and that establishes the second committee.

Exercise 2-4 By axiom 4, the second committee (from exercise 2-3) has three members, so you must add two more members. At least one of them, however, must be a member from the first committee (axiom 2). You might argue that axiom 2 allows the possibility that both these two additions are also in the first committee. To see that this isn't so, see what would happen if it <u>were</u> so, using axiom 1.

Exercise 2-5

| A, B, C | | A, D, E |

| B, D, F | | C, E, F |

Axiom 1 isn't met. For instance, no committees contain A and F, B and E, etc.

Exercise 2-6   Seven committees and seven members:

ABC
DAE
BDF
BEG
CDG
CEF
AFG

Exercise 2-7   No. Before you can use axiom 1, you have to have some members to start with.

Exercise 2-9   <u>Alternate axioms for System 2</u>:

1.   For any pair of points, there is exactly one line containing both of them.

2.   Any two lines have at least one point in common.

3.   There is at least one line.

4.   Every line contains exactly three points.

5.   Not all points are on the same line.

Exercise 2-10   Using axiom 5, we can add one more point <u>off</u> the line, like this:

Now using axiom 1, this fourth point can be joined to the other three, like this:

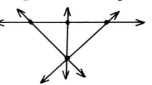

and there are the four lines!

Now axiom 4 can be used again like this:

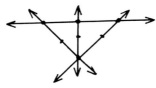

or like this:

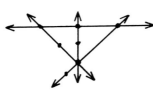

or in some other similar way, and there are the seven points!

Exercise 2-11

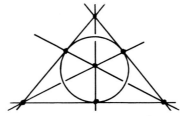

Exercise 2-13   There cannot be more than six members.
There cannot be less than six members.
There are three members on each committee.

Exercise 2-14   There are at least 4 members (exer. 2-3).
There are at least 2 committees (exer. 2-3).
There are at least 5 members (exer. 2-4).
There are exactly 7 members.
There are exactly 7 committees.

Exercise 2-15   same as exer. 2-3, p. 4

Exercise 2-16   two: System 1 and System 2

Exercise 2-17   yes

Exercise 2-18   (1) On page 2, the two sentences beginning "A model doesn't have to agree..."
(2) On page 4, the sentence beginning "When you start with a new set of rules..."
You might also argue that the <u>definition</u> of an axiom on page 6 is a meta-axiom.

Exercise 2-19   No. You cannot use outside knowledge <u>about points</u>, <u>lines</u> and <u>planes</u> (and the way they relate to each other). However, you must use logic and the fact that you understand English, even though this is "outside" knowledge.

Exercise 2-20   Yes, any statement about all meta-axioms would be a meta-meta-axiom. For instance:

Meta-Meta-Axiom 1 In case of a disagreement over the meaning or interpretation of a meta-axiom, that meaning should be chosen which allows the speediest forward progress.

Also, the definition of a meta-axiom is a meta-meta-axiom. A meta-meta-axiom is always an esthetic criterion, not a logical one. Thus math is ultimately grounded in esthetics.

Exercise 2-21   no, no

System 3
<u>System 3</u>

Exercise 3-1 axiom 3

Exercise 3-2 No, see axiom 6.

Exercise 3-4 By axiom 3, there is at least one line:

By axiom 4, the line has three points on it:

By axiom 5, not all points are on the same line:

Call the line above "L" and the point which isn't on it "P" (in anticipation of using axiom 6):

Now by axiom 6, <u>exactly</u> one of the points on L is not joined to P with a line, so one of the three pictures below is true:

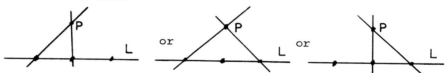

(One doesn't usually work with all three possibilities. Pick one of them, complete your explanation, and then say something like "...and a similar argument can be made for the other possibilities." I will use the one on the left above, though you should convince yourself that the arguments used could equally well apply to the other two.)

By axiom 2, there is another line on (through) P which doesn't intersect L:

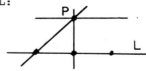

and there are at least four lines, which is the theorem which was to be proved!

Exercise 3-5 (a) Angle A is not acute.
            (b) Line L does not pass through point P.
            (c) A line exists.
            (d) Line L_1 is perpendicular to line L_2.

Exercise 3-7 exercises 1-5 and 1-6, p.3 .

Proof of Theorem 3B: Step 1: Suppose there
is a point that all lines of the system lie on.
Step 2: Draw a picture of this situation.
Step 3: Since each line contains three points,
pick a point on one of the lines which is not
the point that all lines lie on.   Call it P
as shown.   Call one of the other lines L, as
shown.

According to axiom 2, there is one line
through P which has no points in common with
L.   Draw it, as shown.

This new line, having no points in common
with L, cannot pass through the point
that all lines are supposed to lie on.   Since
the axiom cannot be wrong, the supposition
that "there is a point that all lines lie on"
is wrong, and so the theorem is true.   (There
is more than one correct proof for this theorem.)

Exercise 3-8 Hypotheses: (1) It rains.

                         (2) x is a man.

                         (3) Two parallel lines are both interesected
                              by a third line.

      Conclusions: (1) We get wet.

                         (2) x is mortal.

                         (3) The alternate interior angles are equal.

System 4

Proof of Theorem 4A: Suppose two distinct lines can meet in more than
one point:

Now we have two distinct points, A and B, with two lines through them
both.   This contradicts Axiom 1 of Connection.   Hence the supposition
is false so the theorem is true.

Exercise 4-1 axiom 6 (The existence described in axioms 2 and 3 is
                   not unconditional.)

Exercise 4-2 Use axioms 6, 1, and 4.

Exercise 4-3   Axioms 6, 3, and 1 could give the
                following pyramid in which there
                are six lines. The base of the
                pyramid is part of the plane.

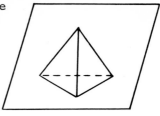

                                        section of a plane

Exercise 4-4   System 2

Exercise 4-5   No, see meta-axiom 1, p.8 .

Exercise 4-6   Neither, we can't tell yet. See meta-axiom 1, p.8 .

Exercise 4-7   axiom 5

Exercise 4-8   No. According to the Completeness Axiom (meta-axiom 4)
                we should be able to tell if statements about the rela-
                tions among objects are true or false. Exercises 4-5
                and 4-6 show that we can't do this yet.

Exercise 4-9   See meta-axiom 3 and Axiom 7 of Connection.

Exercise 4-10   Start with a line:

                By Axiom 2 of Connection, the line
                has at least two points on it.
                Call them A and B:

                By Axiom 1 of Order, there exists
                at least one other point of the
                line between them. Call it C:

                Using Axiom 1 of Order again,
                there is at least one point
                between A and C and at least
                one between C and B. Call them
                D and E, respectively:

                Using Axiom of Order again, there is at least another
                point between A and D, another between D and C, another
                between C and E, etc.

                This process could be continued indefinitely.

Exercise 4-11   In the sketch at the right, line L
                passes through vertex A. We
                would like to allow this
                possibility.

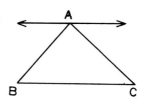

                                                        103

Exercise 4-12 Yes. See the sketch as on the
right. (If a line does inter-
sect a triangle at a vertex,
the axiom has nothing to say.)

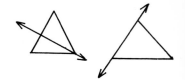

Exercise 4-13 Suppose a line did close on
itself, as shown on the right.
Pick any three points on the
line and call them A, B, and
C, as shown.
Starting at A and moving clock-
wise, C is between A and B.
Starting at A and moving
counterclockwise, B is between
A and C. This contradicts
Axiom 2 of Order.
Therefore, a line cannot
close upon itself.

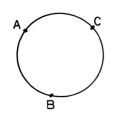

Exercise 4-14

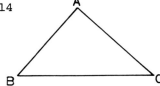

Side BC is opposite vertex A.

Side AC is opposite vertex B.

Side AB is opposite vertex C.

Exercise 4-15 Call the point of intersection P. Choose points A and B
on L$_2$ in such a way that P is between them. Now look at
part (c) of the Axiom of Connection.

Exercise 4-17 Point X is in the interior of angle ABC. Point Y is in
the exterior.
Point C is in the interior of angle RST. Point D is in
the exterior.
Point Q is in the interior of angle MNO. Point P is in
the exterior.

Exercise 4-18 Mentally "block off" line L$_1$ and consider only L$_2$ and L$_3$.
Because of the result of exercise 4-15, L$_3$ must extend
into the "lower" half-plane formed by L$_2$. Now block off
L$_2$ and run a similar argument on L$_1$ and L$_3$, then draw
the necessary conclusion.

Exercise 4-20 GH, NO, O, E

Exercise 4-21 (a) axiom 5   (b) axiom 7   (c) axiom 4
              (d) axiom 8   (e) axiom 1   (f) axiom 3

Exercise 4-22   (a)   Axiom 1 of Connection, p. 16

              (b)   exercise 4-13, p. 20

              (c)   Axiom 1 of Congruence, p. 25

              (d)   Axiom 5 of Congruence

              (e)   Although only three parts of the two triangles are marked as being congruent, they are marked in such a way that Axiom 8 of Congruence (SAS) can be used to say that the triangles themselves are congruent. This means that the other three parts of the triangles are also congruent. (The unmarked parts of congruent triangles are also congruent.)

              (f)   See Axiom 1 of Congruence.

              (g)   Triangle ABC is congruent to triangle BCD because of SAS. Hence the previously unmarked parts are also congruent. In particular, angle DBC is congruent to angle BCA.

Exercise 4-23   (c)   The angles have one side and their vertices in common, but side BC is in the _interior_ of < ABD.

              (d)   Point O is between M and N.

Exercise 4-26   Suppose angles 1 and 2 (as shown on the right) are supplementary and that angle 1 is greater than a right angle. If < 2 were a right angle, then < 1 would also have to be a right angle (definition of perpendicular), which it isn't. Hence < 2 is not a right angle. If angle 2 were greater than a right angle, draw a ray which is ⊥ to L at point O. Then the ray must fall in the interior of < 2. Since < 1 is greater than a right angle, another ray could be drawn perpendicular to L in the interior of < 1 at O. Find the contradiction and finish the proof.

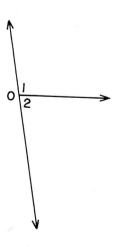

105

Exercise 4-27

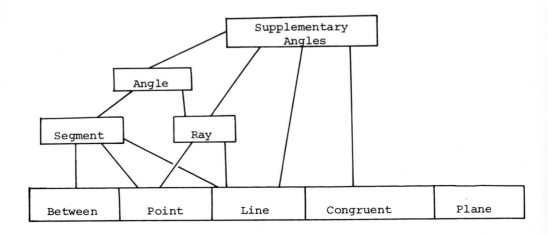

Exercise 4-28

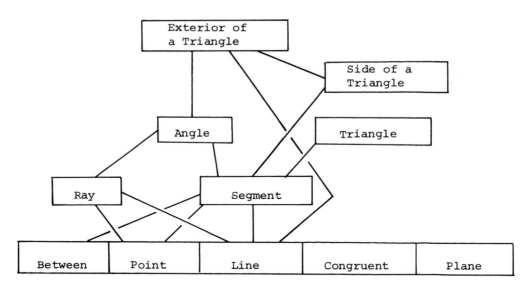

Exercise 4-32   Suppose a triangle has two right
angles, as shown on the right.
Then the exterior angle drawn
at one of the vertices is also
a right angle. Do you see the
contradiction? Similar arguments show that a triangle
cannot have two angles which
are greater than a right angle.

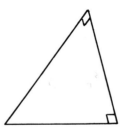

Exercise 5-2   The axioms used in the proof of Theorem 4H (including
those axioms used to prove the theorems which in turn
were used to prove 4H) carry over into System 5. Since
all those axioms are still true in System 5, so is
Theorem 4H. This is not true of Theorem 4J. Do you see
why?

Exercise 5-3   Theorems 4A through 4H carry over into System 5.
Theorems 4J and 4K do not. Don't continue until you
understand why this is so.

Exercise 5-5   No conclusion about such a line is possible.

Exercise 5-7   Yes. This is what the last four pages have been all
about.

Exercise 5-8   Yes. In this case the lines are called "non-intersecting."

Exercise 5-9   Look at the figure on the right,
showing the right- and left-
hand parallels to L through P and
also the perpendicular to L
through P. Beginning with $\overline{PQ}$
and moving your eye counterclockwise around point P, L_1 is the

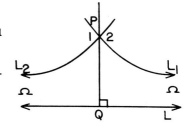

first non-intersecting line on
the right (this is what is meant
by "right-hand parallel through
P.") Starting at PQ and moving
clockwise, L_2 is the first non-

intersecting line on the left.
(This is what is meant by "left-hand parallel through
P.")   Thus all the non-intersecting lines through P
must lie in the interior of angles 1 and 2. If L_1 and

L_2 were really the same line, there would only be one
line through P which does not intersect L, which contradicts the Axiom of Non-Intersection.

Exercise 5-12   The base and summit do not intersect because of Theorem
4H (alternate interior angles are congruent.) By
exercise 5-10, the base and summit cannot be parallel.

Exercise 5-14   Since two right angles comprise a straight angle (definition of a right angle), the two theorems appear to be
contradictory.

Exercise 6-1   Of Theorems 5A through 5K, only 5G and 5H carry over into
               System 6 because the others depend on the Axiom of Non-
               Intersection of System 5. This axiom does not hold in
               System 6.

               Theorems 4A through 4F carry over to System 6, though 4G
               does not (the reason for this will be explained presently)
               Since 4G does not carry over, neither do the theorems
               which follow it and depend on it in System 4: Theorems
               4H, 4J, and 4K.

Exercise 6-3   Let L be a line. Pick a point
               A on L and mark off on L on
               both sides of A a segment of
               length r. Call the endpoints
               P and P'.
               Draw a line perpendicular to L
               at A. Call it L_1. Then P and P'
               are poles of L_1. L and L_1 inter-

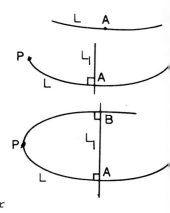

               sect at another point. Call it B.
               Extend L past B. It must pass
               again through P'. Continue to
               extend L past P' a distance r and
               call the endpoint C.
               C must be on L, for suppose not.
               Draw a line through C perpendicular
               to L. Call it L_2. Let D be the

               point of intersection of L and L_2. Connect P' to D and
               arrive at a contradiction. Since C is on L it must in
               reality be the point A and L has length 4r.

Exercise 6-5   Draw a segment from a vertex perpendicular to the opposite
               side if the triangle has three acute angles. Draw it
               from the vertex of the right angle if the triangle has
               one. In either case, the segment must intersect the
               opposite side at a point between the endpoints of the side
               (suppose not). Then use Theorem 6G.

               NOTES ON UNCOMPLETED PROOFS

System 4

Theorem 4B   By axiom 4, there is exactly one plane which contains
             points A, R, and B. Since A and R are in this plane, axiom
             5 says all of L_1 is in it. Since B and R are in this plane,
             axiom 5 says all of L_2 is in it. So L_1 and L_2 are both in
             the same plane, as was to be concluded.

Theorem 4C Since $\overline{YW}$ lies in the interior of $<XYZ$, $<XYW \lessdot <XYZ$, hence we have a contradiction, and so $\overline{AC}$ is not smaller than $\overline{XZ}$. A similar argument shows that $\overline{AC}$ is not larger than $\overline{XZ}$, so by the Axiom of Trichotomy, $\overline{AC} \simeq \overline{XZ}$. Thus the triangles are congruent by SAS.

Theorem 4G Since $\overline{DA}$ lies in the interior of <4, <4 is larger than $<MAD$ and hence larger than <1.

To prove that $<4 > <2$, draw a segment from B to the midpoint of $\overline{AC}$ (call the midpoint N). Extend this segment past N to a point E in such a way that $\overline{BN} \simeq \overline{NE}$. Now draw $\overline{AE}$ and finish the proof.

Theorem 4J Suppose in the figure on the right, that $<1 > <2$. Through the vertex of <1, draw a third line L_4 in such a way that it makes an angle with L_3 which is congruent to <2 (Axiom 5 of Congruence). Now use Theorem 4H and the Axiom of Parallelism to reach a contradiction.

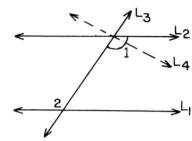

Theorem 4K Draw a triangle, as shown on the right. Through any vertex, draw a line parallel to the opposite side (Axiom of Parallelism). Then use the previous theorem to finish the proof.

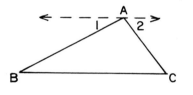

System 5

Theorem 5B Suppose there is a last intersecting line on the right which intersects L at a point R. Pick a point on L to the right of R, join it to P, and find the contradiction.

Theorem 5C Since $\overline{SP}$ is in the interior of <2, $<SPQ$ must be smaller than <2. From the last two sentences of the proof given on page 60, $<SPQ \simeq <2$. This is a contradiction. Hence <1 is not greater than <2. A similar argument proves that <2 is not greater than <1. Hence by the Axiom of Trichotomy, $<1 \simeq <2$.

Theorem 5D Suppose they were right angles, as shown on the right. Now look at the Axiom of Perpendicularity and obtain a contradiction.

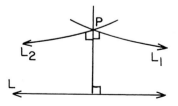

Suppose they were obtuse, as
shown in this figure.
Since < 4 must be acute, so
must < 5, so that the left-
hand parallel would lie in
the interior of the right-
hand angle of parallelism.
Find the contradiction and
finish the proof.

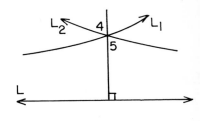

Theorem 5G   In the figure on the right,
draw $\overline{AM}$ and $\overline{BM}$. Then $\triangle$ AMC
$\simeq$ $\triangle$ BDM by SAS. Hence $\overline{AM}$
$\simeq \overline{BM}$ so that $\triangle$ AMN $\simeq$ $\triangle$ BMN.
Thus < ANM $\simeq$ < BNM. Also
< AMN $\simeq$ < BMN and < AMC $\simeq$
< BMD. Now use the defini-
tion of perpendicular to
finish the proof.

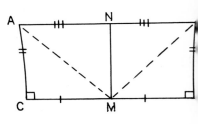

Theorem 5H   In the figure for the proof of Theorem 5G, < CAM $\simeq$ < DBM
and < MAN $\simeq$ < MBN. Hence < CAN $\simeq$ < DBN (suppose not!),
as was to be concluded.

System 6

Theorem 6J   If the two equal sides weren't less than r, they would
intersect. Then use Theorem 6H to explain why the base a▮
summit must be less than 2r.

Theorem 6L   Case 2: The perpendicular from C to $\overline{AB}$ must intersect $\overline{AB}$
between A and B (suppose not) forming two right triangles
This perpendicular may or may not divide the third angle
into two acute angles. Discuss the result either way.

Index

angle(s), 22
  acute, 58
  alternate-interior, 34
  exterior of, 23
  included, 24
  interior of, 23
  obtuse, 58
  of parallelism, 59
  right, 26
  straight, 34
  supplementary, 34
  vertical, 34
axiom, 3, 6
  set, 6

base
  isosceles triangle, 45
  Saccheri quadrilateral, 64
Betweenness, Axiom of, 79
bisect
  an angle, 45
  a segment, 45

circle
  area, 96
  circumference, 96
completeness, 18, 85
conclusion, 13
Congruence, Axioms of
  Euclidean geometry, 25-26, 39
  Lobachevskian geometry, 50
  Riemannian geometry, 72
congruent, 19, 36
Connection, Axioms of
  Euclidean geometry, 16-17, 38
  Lobachevskian geometry, 49
  Riemannian geometry, 71
consistency, 47, 84
converse, 42

Dedekind's Axiom, 20, 38, 50
deduction, 8
deductive system, 3, 7
definition, 22, 35-38
denial, 11
distance, 78
  polar, 78

elliptic geometry, 71
endpoint, 19
Euclid, 16
Euclidean geometry, 16
existence rule, 2

formulas, 96
Fundamental Elliptic Axiom, 73

Hilbert, David, 16
hyperbolic geometry, 67
hypothesis, 13

implication, 13
independence, 47, 85
Intersection, Axiom of, 79

length, 72
line, 16, 36, 49
Lobachevskian geometry, 48, 67
logic, 13, 42

Measure Axiom, 72
meta-axiom, 7
meta-axiom 1, 8
meta-axiom 2, 11
meta-axiom 3, 16
meta-axiom 4, 18
meta-axiom 5, 33
meta-axiom 6, 47
meta-axiom 7, 47
midpoint, 19
model, 2

non-Euclidean geometry, 48
Non-Intersection, Axiom of, 51

objects of a system, 7
Order, Axioms of
  Euclidean geometry, 19, 38
  Lobachevskian geometry, 49
  Riemannian geometry, 71

parallel, 35
  left-hand, 56
  right-hand, 56
  through a point, 56
parallelism, angle of, 59
Parallelism, Axiom of, 33, 39, 48
parallelogram, 46
  area, 97
perpendicular, 26, 35-36
Perpendicularity, Axiom of, 79
plane, 16, 36, 49
Playfair's Axiom, 33
point, 16, 36, 49
pole of a line, 77
primitives, 36
proof, 7, 8, 13
  indirect, 11-12
proof rules, 30-31
proportion, 86
Pythagoras, 92
Pythagorean Theorem, 92

quadrilateral, 46
  Lambert, 67
  Saccheri, 64, 82-83

ratio, 86
ray, 22
rectangle
  area, 96
  perimeter, 96

relations of a system, 7
Riemann, Bernard, 71
Riemannian geometry, 71

segment, 19
Separation, Axioms of
  Euclidean geometry, 21, 38
  Lobachevskian geometry, 50
  Riemannian geometry, 72
square
  area, 96
square roots, 91
summit, 64
summit angles, 64
system, deductive, 3, 7

theorem, 6, 13
  proof of, 7
transitivity, 25, 26
trapezoid, 46
  area, 97
triangle(s), 19
  area, 97
  congruent, 25
  equilateral, 44
  exterior angle of, 34
  infinite, 57
  isosceles, 44
  similar, 88
Trichotomy, Axioms of
  Euclidean geometry, 30, 39
  Lobachevskian geometry, 51
  Riemannian geometry, 73

undefined terms, 36

vertex of a triangle, 19
vertex angle, 45

32-103